Tantalisingly Close

MediaMatters is a series published by Amsterdam University Press on current debates about media technology and practices. International scholars critically analyze and theorize the materiality and performativity, as well as spatial practices of screen media in contributions that engage with today's digital media culture.

For more information about the series, please visit: www.aup.nl

Tantalisingly Close

An Archaeology of Communication Desires in Discourses of Mobile Wireless Media

Imar O. de Vries

Amsterdam University Press

This book is published in print and online through the online OAPEN library (www.oapen.org).

OAPEN (Open Access Publishing in European Networks) is a collaborative initiative to develop and implement a sustainable Open Access publication model for academic books in the Humanities and Social Sciences. The OAPEN Library aims to improve the visibility and usability of high quality academic research by aggregating peer reviewed Open Access publications from across Europe.

Cover design: Suzan Beijer, Amersfoort
Lay-out: JAPES, Amsterdam

ISBN 978 90 8964 354 4
e-ISBN 978 90 4851 491 5
NUR 670

Contents

Preface and acknowledgements

The following is a true story, as related to me by a student in my Masters course on wireless communication. This student (whose identity I will protect by calling him Oliver) had had a big night out on the town. At three in the morning he and a flatmate decided to head home, only to discover at their front door that they both had forgotten to bring their keys with them. There was no need to worry, they thought; Oliver quickly took out his mobile telephone from his pocket, and dialled the number of their one flatmate who had stayed home. Much to their dismay, however, their sleeping friend had no desire whatsoever to be disturbed at that late an hour, and had shut off his mobile telephone. Dumbstruck by this unexpected setback, Oliver could think of only one solution for the awkward situation he and his friend now both found themselves in: get on their bikes and cycle twenty kilometres out of town to his parents' house to get a spare key. Once there, however, it turned out that his parents were away for the weekend. It was only on their way back home, exhausted by their nightly workout and still wondering how they could get into their house, that it finally dawned on Oliver. There had been a much easier way to signal his flatmate to have him open the door than trying to call him on his mobile... the doorbell.

At the beginning of the twenty-first century, mobile communication devices are everywhere. What is more, they have quickly come to be considered – sometimes foolishly so – as essential tools to handle almost any problem or situation in life. It is astonishing to realise that merely two decades ago only very rich business people could afford them, and that owners of mobile telephones were regularly being mocked for thinking that it was necessary to be in constant touch. I can still distinctly remember a Dutch comedian sneering, when he ridiculed the rising use of mobile car phones in 1992, that truly important persons were supposed to be unreachable. Today, his point arguably still stands, but the mobile telephone has long ceased to be the rare and pompous status symbol of the wealthy. Moreover, it has been joined by countless other communication technologies that keep people and objects inconspicuously connected. We now live in a world in which Blackberrys, WiFi-enabled laptops, iPhones, RFID chips, Android smartphones, and so forth, have all but completely pervaded everyday communication processes, making many people wonder how they could ever have done without them. It is at times such as these that we have to be aware that mobile communication devices, in their mundaneness, become almost invisible, as they come to constitute our living environment. Herein lies the opportunity for the observant media scholar to intervene, and expose what is actually happening... and why.

I was still studying to become a media scholar when, at the end of the 1990s, I got my first mobile telephone. It was an Ericsson A1018s, a light-blue device measuring around thirteen centimetres long, five centimetres wide and three centimetres deep, with a big antenna dangerously protruding from the top. It could make mobile calls and send and receive text messages, but that was it. No games, no camera, no polyphonic ringtones, no MP3 player, no video capabilities, and certainly no Internet or apps. By today's standards it would be considered a bulky and feature-poor device, but at that time I was as happy as could be, very much aware of my newly acquired position in a rapidly changing communication landscape. And it happened quite by accident, actually. Although I had already been considering buying a mobile telephone – especially after my grandparents had beaten me to it that summer, having purchased one for use in their rural holiday house in France – I was waiting for just the right opportunity, model, and price. However, I ended up acquiring one in a rather unconventional way: when my parents bought themselves a new bed, they received with it a complete mobile telephone package for free. As my father already had a car phone and my mother saw no use for it, they decided to give it to me. Without having to think twice about it, I happily accepted the device.

I still have that first mobile telephone. It is kept in a drawer, next to the Sony Ericsson T65i, K700i, Z600, and K800i that I have owned in the past couple of years. They are tangible evidence of not only my personal preference for Sony Ericsson mobile telephones, but also of the pace with which the devices have evolved, boasting more and more features with every generation. On a very small scale, my collection represents what for many may seem to be a natural process in the development of communication technologies: they become better every time, increasingly fulfilling our communication desires. This highly charged idea of communicative progress is what has fascinated me ever since I began to study media, and it fascinates me still. It has led me to write this book, and I expect it will provide me with many more projects, as it will continue to influence media development. Every time someone asks whether I fear my work on mobile communication devices might soon become outdated because of the rapid rate of new product introductions, I smile, and say that each press release touting the long-awaited arrival of the world's best communication device actually reconfirms what I firmly believe: like the torment of Tantalus, we are convinced that if only we try harder, we will be able to reach far enough and finally fulfil our desires. And yet, we always fail.

This book, then, is about communication utopia. It is about what could be considered a tragicomical quest for communicative fulfilment, a quest that has manifested itself countless times in human history, and that will likely continue to be part of us forever. Mobile communication devices have become the latest embodiments of that quest, and probably more literally than one might think; in every single device, you will find capacitors that are made with a heavy metal called tantalum. So, each time you use one of them to fulfil your desire for communication, remember: a true and pure connection is, and always will be, tantalisingly close.

There are many people I would like to thank for helping me in my own personal quest to finish this book. It has been a long and winding road, but never a lonely one. First, I must thank the people at the Faculty of Humanities, at the Research Institute for Culture and History, and at the Department for Media and Culture Studies at Utrecht University for creating the best job in the world; it truly is wonderful and inspirational working within the New Media & Digital Culture (NMDC) Masters programme.

Ah, the NMDC group! Could there be a friendlier, funnier, more creative, intelligent, and highly entertaining bunch of people than those that made me feel so much at home at Utrecht University? Our irregular meetings in classrooms, restaurants, conference halls, and homes have always been truly pleasurable and exciting, and I hope to be part of many more. The group would not be complete without its mafiosi, though; my utmost gratitude goes out to Marianne van den Boomen and Mirko Tobias Schäfer, without whom my academic and personal life would not have been enriched with so many memorable experiences. Our seminars, trips to faraway places, nights out on the town, heated debates, revelatory and consolatory encounters, and, above all, our discussions on our research projects, have all contributed to what I consider to be the best stock-in-trade a person can have on the route towards academic achievement. May we continue our fabulous friendship for many years to come.

There are more who helped me by giving feedback on my ideas, writing style, English vocabulary and grammar, and intermittent chapters; a big thanks to Andreas Fickers, Steph Harmon, Jim Hurley, Frank Jansen, Tim Muentzer, Mathijs Sterk, Merijn van der Vliet, all the participants of the Utrecht media seminars, and all the students who attended my two Masters courses 'Wireless Future' and 'New Media Archaeology'. Furthermore, I am deeply grateful to Simone Veld, for helping me regain my motivation when I most needed it.

A special word of recognition and appreciation must be given to William Uricchio and John Durham Peters, who, through their lectures, articles, and books, have shaped my academic mind and have given me the basic building blocks with which I was able to develop my own research interests. I hope this book does justice to your work. Every bit as important in this respect are my colleagues Frank Kessler and Joost Raessens, who provided me with direction and valuable knowledge, were always willing to expand on things as necessary, and kept encouraging me to be creative yet precise in my thinking.

A further acknowledgement goes to my friends who have given me moral support during the many months when I hardly had the time for a social life: Karin Algera, Daniëlle Berends, Rudi de Boer, Ernst ten Bosch, Ildikó Flesch, Martine Gusky, Gouke Hilte, Fried Hoeben, Marije Korver, Jurjen Kramer, Wouter Kroese, Inge Judith Nobel, Sander van Ouwerkerk, Petra Overdevest, Bart Sassen, Marjolein Stikker, Arjen Veenhuizen, Vincent van der Velde, Raymond Viveen, and Simon Welling. Thanks as well to Dino, whose fur therapy often provided a very effective guard against mental stress.

My sister and my parents deserve as much credit as I can possibly give; their unconditional love and unfaltering support have brought me to where I am today.

And finally, I offer my eternal gratitude and love to my wife, Kim Raad, who was always patient, understanding, and never stopped believing in me. This is for you, my dearest.

Utrecht, August 2011

Introduction

By connecting all kinds of objects everywhere, we aim to achieve genuinely ubiquitous communications. To this end we are developing fourth-generation technologies to realize a true ubiquitous society. (NTT DoCoMo 2006)

Envisioning the mobile future

In 1999, Japanese mobile communications giant NTT DoCoMo created Vision 2010, a corporate strategy aimed at countering the effects of a slowing growth of the market at the time. The project was framed in a familiar, tried-and-tested manner: by portraying the contemporary state of mobile communications as one that served user demands well but would inevitably run into problems with an expected increase in the consumption of information, a necessity of progress towards a better future presented itself. In a press release that announced its lofty intentions, the company clearly was not afraid to allude to a command over supernatural powers in bringing about this better future, as it promised to improve 'the quality of life' by use of MAGIC: 'Mobile multimedia; Anytime, anywhere, anyone; Global mobility support; Integrated wireless solution; Customized personal service' (NTT DoCoMo 1999). The use of fanciful acronyms did not stop there; the goal of 'Vision 2010' was to be achieved by following the company's 'DREAM': Dynamics, Relationship, Ecology, Action, and Multi-view, all words that were elaborately described as being in some way connected to ideas of change, growth, collaboration, and progress (ibid.). The message was unambiguous, and has remained so up until the present day: as the world progresses, new and ubiquitous mobile communication technologies will increasingly help to fulfil our communicative needs – all thanks to NTT DoCoMo of course.

To promote Vision 2010, and thus make visible the company's ideas of how their new wireless technologies would help shape the future of communications, NTT DoCoMo over the years created a series of concept videos that aimed to unveil the world 'beyond the mobile frontier'. Each video shows people using wireless communication technologies in a wide range of everyday and less common situations, and gives an impression of how these technologies serve to make life more convenient. In one of the first videos that was released, we are introduced to 'a typical Japanese family of four' and are given an insight into their lives both in the very near future (which, at the time the video was made, was projected onto the year 2003) and in the more distant future (set in 2010).[1] In the first part, we see the family members use their mobile devices to audiovisually communicate with each other wherever they are, take remote English classes outside in the park, control home appliances from a distance, receive offers from nearby stores, download music, pay for cans of soda and connect to office databases from the back seat of a moving car. In the second, more futuristic part, they partake in virtual three-dimensional academic courses in which all spoken languages are automatically translated, use jewellery as user identification systems that communicate with coffee machines, tell virtual agents to help plan dinners, and talk to holographically projected friends. All the while and at regular intervals, a male voice-over sums up the technologies and services with which all these activities have been made possible, stressing each time that they *will* happen and *will* become commonplace.[2]

What is important to consider when analysing what NTT DoCoMo's strategy directors think is in store for us is not so much the question whether their short-term predictions have come true or if their long-term predictions are composed of fantastic nonsense. It is far more interesting to study the language and images that are used to convey ideas of what communication technologies are for. The videos not only represent the discourse of desire that is a defining aspect of so many advertisements, but they are exemplary of the way in which telecom industries in particular use a very specific notion of the place that mobile communication devices take in the overall media landscape. Additional textual analyses of the NTT DoCoMo videos would show that they stress the potential gains to be made in productivity, sociability, efficiency, safety, reassurance, and mutual understanding; in other words, they stress communication improvement and its benefits, and present wireless communication technologies as the ultimate facilitators. In many more commercial advertisements, press releases, and articulations of popular culture, a similar reasoning process can be distinguished, one that is repeated as a mantra: mobile communication devices are the self-evident technological expressions of a natural progression in the quest for ever-improved communication.

At first, this observation might seem a bit stretched. Is it not simply a shrewd, yet obvious, marketing strategy to present merchandise in such a way that people think they need it in order to enhance their lives? Are mobile communication devices not just another product to sell? The quick and straightforward answer to both questions would be yes, of course. People buy mobile telephones just as they buy other consumer devices; even if they still function quite well, they are effortlessly replaced by new versions that feature the latest gadgetry. Without advertisements continuously telling us that we need to upgrade to the latest smartphone the pace of adoption might not be as fast as it is now. But think of this: is today's mobile-saturated world truly merely the result of clever marketing? Have we all been fooled by commercial advertisements into buying mobile telephones, and would this never have happened if we had just not bought into all the marketing hype? The answer is a resounding no, there is definitely more to it than that. The idea of personalised mobile communication carries with it an immensely powerful lure that is not simply pushed into our susceptible brains. By portraying mobile communication devices as sources of unlimited potential that can solve any problem imaginable while completing any known lifestyle, advertisements certainly paint a pretty picture, but they also readily tap into our deepest desires for wholeness and completeness, which we hope can be fulfilled by improving communication.

This brings us to the task of figuring out what role our desires for communication play in bringing about what can be called the feverish mobile communication condition of today. Before we can do so, however, it is necessary to ask what exactly that fever in the current mobile communication condition is.

Hark now, hear the mobiles ring!

For quite some time now, mobile communication devices have been heralded as revolutionising progress and radically changing for the better the ways we communicate, do business, think of time and space, play, romance, arrange meetings, and so forth. And indeed, the speed and vigour with which these small computers have pervaded many of our lives is astounding, not to mention the impact they have had and continue to have on the way communication is conceived today. In order to illustrate how this process is typically captured in numbers, here are some examples: in October 2010, the International Telecommunication Union (ITU) reported that in the developed world mobile penetration was at an estimated 116.1 percent, which is to say that there were more subscriptions than inhabitants (ITU 2010b); in January 2011, Informa Telecoms & Media predicted that SMS text messages sent in 2015 will reach a worldwide total of no less than 8.7 trillion, having risen from 5 trillion in 2010, making it one of the most accepted and popular communication channels in the world (ITM 2011); the GSM standard, first commercially launched in Finland in 1991, took 12 years to reach one billion subscribers and just a further two and a half years to reach the second billion (GSM Association 2006).

These are impressive numbers, and, although they do not portray a balanced perspective on how mobile communication devices are actually incorporated into our daily lives, they are more often than not presented as the definitive proof of a process of progress that is impossible to ignore. The perception that the arrival of commercially viable mobile communication devices has been sudden, and the fact that their penetration rate is the fastest of any personal electronic consumer good in history, make a good case for the temptation to attribute revolutionary characteristics to them. And how could they not be revolutionary, the industry might argue. In the third quarter of 2010 the ITU estimated there would be a staggering 5.3 billion mobile telephone subscriptions at the end of the year (ITU 2010a), and there are certainly more people who have access, sharing phones among family members and friends or borrowing them temporarily from savvy businessmen.[3] Even non-owners who actively seek to avoid using the devices can hardly escape their presence in public space. Mobile communication devices are ubiquitous, and despite the fact that they appear in countless sizes and forms, boasting the latest technical tricks or simply providing just talk and text messages, there is a certain characteristic presupposed in them, representing what has become the quintessential symbol of communication in the twenty-first century: to provide a uniquely identifiable wireless connection, in whatever form, to any other node in the information network that is our global society today.

This hype of celebrating a seemingly endless potential found in the sheer abundance of mobile communication devices can easily lead to oversimplified and deterministic notions that the possibility to own or access such a device apparently does not discriminate between socially constructed categories such as class, race, age, ethnicity, gender, or social identity.[4] This, in turn, has resulted in high expectations that the mobile telephone can – finally – provide realisable solutions to long-time political, economic and social problems, again highlighting its revolutionary utopian potential. These expectations are visible in multiple layers of society: not only in popular press, specialised magazines, and science fiction, but also in academic analyses of changing communicative practices and political manifestos. Not to forget, of course, jubilant press releases by the mobile industry. In a telling example of this last category, on 13 June 2006, a few days before the second billionth GSM mobile telephone user was to be connected, Rob Conway, then CEO of the GSM Association (GSMA), said that 'GSM [had] become the first communications technology to have more users in the developing world than the developed world', which meant that 'mobile phones are "bridging the digital divide" at an astonishing rate with relevant, affordable solutions that help families stay in touch, businesses to grow and economies to develop' (GSM Association 2006). Further initiatives taken under the flag of GSMA's vision 'Connect the Unconnected' were to develop a 'sub $30 low-cost mobile phone' and to call for the 'removal of tax barriers on mobile products and services in many emerging countries' (ibid.).

Admittedly, expectations proclaimed by highly placed industry executives can occasionally become reality, for instance when the application of strategic planning reveals typical examples of self-fulfilling prophecies.[5] In the case of Conway's assertion that the digital divide will be bridged by mobile communication devices, the prophecy does appear to come true: sometimes called a leapfrogging technique because it avoids the expense of laying land lines, the adoption of wireless technologies by developing countries has indeed resulted in millions of people having access to communication and information, which they could not obtain before. Moreover, and building on the argument that gaining access alone does not automatically entail social or economic equality, several studies conducted by Jonathan Donner, researcher in Microsoft's Technology for Emerging

Markets Group, have shown that the introduction and use of the mobile telephone in countries such as Rwanda have a significant measurable effect on social and business networks (Donner 2003; 2005).

However, self-fulfilling prophesies only work when they account for technical and economic realities, and more often than not facts of reality are bent in order to suit a dominant paradigm in culturally constructed discourses. There is, for instance, the most notorious and controversial example of the supposedly democratic potential of the mobile telephone, that of the ousting of Philippine president Joseph Estrada in January 2001, which was thought to be brought about by thousands of Filipinos who used text messages and phone trees to organise mass protests.[6] This illustration of the mobile telephone's assumed ability to facilitate political change pops up in many an analysis of its postulated far-reaching potential; it has become a stock argument that confirms and sustains myths of the power of wireless technology. It is a truism to say that the mobile telephone does not inevitably oppose dictatorships and help construct or restore social cohesion,[7] but what is important to note here is that high expectations, whether they are fulfilled or not, *do* have a considerable impact on how the discourse of mobile communication technology is moulded, and how an aura of utopian possibilities is maintained.

Media technology's seductive promises

The utopian aura is but one aspect of mobile communication discourse, and cannot in itself explain why wireless technologies in particular attract so much popular anticipation and scholarly attention. However unprecedented the potential of mobile communication devices actually is to accomplish social, political, and economic equality, the gush of revolutionary and utopian claims that accompanies their incessant rise is not unique. Every existing communication technology was once new and full of promise, as many scholars have demonstrated. Wireless telegraphy, at its inception, was seen as 'the means to instantaneous free communication' (Flichy 1995: 109); telephony seemed to promise the banishment of distance, isolation, and prejudice (Briggs 1977: 45); radio was poised to pave the way for contact with the dead (Sconce 2000: 59-61); and television would transform its viewers into all-knowing eyewitnesses of everything that went on in the world (Elsner, Müller & Spangenberg 1994: 110). However, all such claims have, in one way or another, been condemned as much as praised (Davis 1976), or proved inaccurate. Most of the exuberant expectations never really came to fruition, many media technologies have developed in a different way from what was foreseen, and no single medium has been able to completely replace all other media. Nevertheless, the process of praise for unprecedented opportunities remains conspicuous throughout media history.[8]

This observation raises two important and interconnected issues regarding how to understand today's frenzied mobile communication condition: the first is that, in spite of reality continuously disarming them, utopian and revolutionary claims appear to be part and parcel of the arrival of new technologies; the second is that these claims apparently possess and share particular discursive characteristics. Let me briefly expand on this. The first issue, of the recurrent nature of unbridled reverence for new media, has been addressed only relatively recently in studies of media history. It is in Carolyn Marvin's insightful *When old media were new* (1988), for instance, that we read of how telephones and electric lights – new technologies at the end of the nineteenth century – were not only awaited with anxiety but were bestowed with magical beneficial powers as well. Since, in a historical sense, new technologies have 'simply been elaborations' on old technologies, Marvin argues, it should not come as a surprise that our expectations are therefore quite alike (Marvin 1988: 3). The same point is made by Vincent Mosco in *The digital sublime* (2004), when, presenting us with a historical overview of utopian hopes and desires in media reception, he tells that '[p]ractically every substantial technological change has been accompanied by similar claims' (Mosco 2004: 119).

While Marvin does not fully explore in detail the possible explanations for the existence of these 'hymns to progress', Mosco holds that such hopeful stories are 'myths' that celebrate the 'technological sublime' (ibid.: 22). By referring to the sublime as a powerful trigger to provoke the creation and telling of myths that tout technologies, Mosco here explicitly sides with a range of scholars that have done so before him, or have used a similar notion of how the sublime engenders rhetorics of progress. Coined by Perry Miller (1961) and more fully developed by Leo Marx (1964) and David E. Nye (1994), the term 'technological sublime' not only alludes to an 'essentially religious feeling' of complete and transcendent excellence induced by the confrontation with impressive technology, but also is a 'defining ideal' that binds people together (Nye 1994: xiii-xiv). As such, the sublime is both an experience and an orienting force, tapping into 'fundamental hopes and fears' and representing 'a way to reinvest the landscape and the works of men [*sic*] with transcendent significance' (ibid.: xiii). It can be identified on many different levels; as Nye has noted, the last decades of the nineteenth century for instance saw the rise of what could be called the 'electrical sublime', engendering and composed of a widespread belief in the supposedly unlimited powers of man-made electrical contraptions (Nye 1994: 143-172).

Myths play an important part in upholding the sublime as an orienting force, Mosco maintains, as they

> mean more than falsehoods or cons; indeed, they matter greatly. Myths are stories that animate individuals and societies by providing paths to transcendence that lift people out of the banality of everyday life. They offer an entrance

> to another reality, a reality once characterized by the promise of the sublime. (Mosco 2004: 3)

Myths, as culturally naturalised ideas, offer captivating promises that are 'unfulfilled or even unfulfillable', which makes them seductive and powerful (ibid.: 22). Without so much saying so in such terms, Mosco defends the view that we live with an instinctual urge to be astounded, that we are equipped with a persistent willingness to believe revolutionary and utopian narratives because of the lure of unfulfilled potential in new technologies, even if we, paradoxically, already know we will never be able to actually reach the complete and transcendent excellence that is the sublime.

This brings us to the second issue that needs to be addressed before we can take on the investigation of the mobile communication condition, which is the one that concerns the particular characteristics of discourses of new technologies that bring about the continuous recurrence of their idealisation. If we are indeed able to observe that utopian expectations not only are not unique to mobile communication devices, but can also never be met completely, then surely there is something that makes us (want to) forget this, something that coerces us to project high hopes on the new communication technologies anyway. Here, the reason lies in the adjective used in the term 'new technologies': arguably, nothing is as seductive and full of promise as the phenomenon of 'new' itself. As philosopher and media theorist Boris Groys (2002) notes, the 'new' has traditionally been connected to 'the idea of progress, the utopian future': it lures, it does away with what was before, and replaces things, often with a shock or with profound consequences. It signifies change and difference, processes that are often associated with improvement; in many an industry's jargon, 'innovation' is a must-use word. The new can also mark the endpoint of a circular process, and announce the start of another iteration, or better yet, another opportunity. A new moon or a new day is not necessarily an improvement over the previous version, but forms the next iteration of a continuous loop. Thus, the new transforms but also allows repetition, and therefore offers promises of new hope and new chances. As such, it contributes to the persistent feeling that, whenever something new presents itself, opportunities to come closer to or to return to a sublime state of being are near.

Questioning the communication sublime

What lies at the heart of this book is the intent to grasp the functioning of one of the strongest and most appealing recurrent utopian myths that accompanies the advent of new media, which is that communication technologies will eventually endow us with omnipresence and omniscience. The expressions of this myth are given shape by what I call the 'communication sublime', an awe-inspiring and immensely tantalising vision of a final and universally accessible communication space where the accumulation and dissemination of information stands for the most important condition of human progress, and where there can be no misunderstanding. Manifestations of this myth are visible in stories of success as well as in those of failures, in press coverage, research reports, commercial advertisements, and popular culture. What makes a study of this myth especially interesting today is that wireless technology discourse upholds the illusion that mobile communication devices come closer to realising the myth's expectations than any other medium before. The 'anywhere, anytime, anyhow, anyone' slogan, both obviously and subliminally attached to every mobile apparatus, has convinced countless numbers of people that a mobile communication device is of essential importance to make their lives complete. Moreover, the record-breaking pace with which mobile communication devices have spread among a multitude of cultures suggests that they fit a fundamental communicative need, one that is globally felt. As James E. Katz and Mark Aakhus concur in the introduction of their aptly titled book, *Perpetual contact* (2002):

> In the history of human imagination the power of real-time interactive oral communication over great distances had been a power so great that even most divine beings were considered incapable of it: Zeus, king of the Greek gods, and the rest of the pantheon, had to rely on messenger-boy Mercury. Today a good many messenger boys have their own mobile phones. (Katz & Aakhus 2002: 2)

By referring to mobile telephones and equating their functioning to the powers of the supreme master of communication, Mercury,[9] it is clear that Katz and Aakhus ponder the idea that this god's capabilities are now literally within our grasp, and that mobile communication devices are the outcome of an elaborated process of looking to fulfil a long-standing human communication dream.

Like Katz and Aakhus, I see an overarching logic associated with the evolution of communication technology, one that is 'grounded in broader ideologies that are rooted in historical, materialistic, religious, and ideational ontologies, and that have dominated human attempts to contextualize and make meaningful their life experiences' (ibid.: 305). In other words, when we look at how communication technologies have historically been imagined, constructed, received, and criticised, we may be able to uncover a continuously present undercurrent, a recurring discursive trope that holds that new communication technologies will improve our lives as long as they bring us closer to reaching the communication sublime. Katz and Aakhus are on a similar track: they trace the origins of such an undercurrent in the compelling image of 'perpetual contact', a myth that presents us with 'an idealization of communication committed to the prospect of sharing one's mind with another, like the talk of angels that occurs without the constraints of the body' (ibid.: 307). The ideal of angelic communication is but one expression – albeit a powerful one – of the communication sublime, but what all of the myths share is a seductive power, one that deceivingly beckons to us with an ultimate form of communicating that awes, inspires, terrifies, and amazes, but that can never be truly reached.

Tying together the observations on the mobile communication hype, the presence of myths in media evolution, and the haunting characteristics of the communication sublime, this book aims to answer the question in what ways the ongoing evolution of communication media is oriented by recurrent myths that give expression to inherently unfulfillable desires to arrive at the communication sublime. We will see that, throughout time, the communication sublime has known many articulations through many myths and has expressed itself in different discursive formations with each new attempt to improve communication. Mobile communication technology is but the latest striving to realise a communication heaven on earth, and although it has opened up new paths that undisputedly lead towards an affirmation of the 'anywhere, anytime, anyhow, anyone' mantra that is a part of the communication sublime, just as with other, older media it cannot prevent misunderstanding. In other words, the communication paradox is upheld: knowingly unable to actually reach any of the idealised ideas of communication, we still keep on trying, thinking that as long as we solve problems of communication by inventing new and better technologies, we will eventually eliminate all obstacles that stand in the way of true and harmonious understanding.

How to approach communication myths in media history... and why

The defining role of the new in fuelling myths, sustaining the communication sublime and giving rise to a déjà vu of utopian claims, has had little attention in the field of media history up until the 1970s. In earlier studies, media history is typically seen as a chain of successful breakthroughs brought about by great inventive geniuses, in which 'newness' merely functions as a marker on a teleologically constructed timescale, indicating how far we have progressed from primitive machinery to cutting-edge state-of-the-art technology. As Siegfried Zielinski writes, such genealogies are 'comforting fables about a bright future, where everything that ever existed is subjugated to the notion of technology as a power to "banish fear" and a "universal driving force"' (Zielinski 2006a: 3). In the last decades of the twentieth century, however, sociocultural studies have demonstrated the fallacy of such a deterministic vision by showing that failures can be just as revealing as successes (Bijker 1992), that media do not simply replace each other but are concurrently present in a broad media spectrum (Flichy 1995), that media never remain unchanged nor are they used solely for intended purposes

(Bolter & Grusin 1999), and that there are chance discoveries, power struggles and social necessities that influence technology development (Winston 1998). Still, the myth that we move forward along a line of successive and successful media improvements persists, not in the least of course in strategies of commercial industries that thrive on maintaining a sense of non-fulfilment that continuously needs to be addressed.

In such progressive accounts of development, media are invariably characterised as having ontological states instead of as being procedurally induced unstable phenomena that constantly change appearance depending on the context in which they are viewed.[10] While the procedural view is far more elegant and takes into account the multitude of factors that influence media development, it is the ontological view that perfectly fits the pragmatic longing for cause-and-effect explanations of why our media are what they are, and what they will become in the future. In order to consider the functioning of both views, I aim to implement the two guidelines of what media historian Erkki Huhtamo proposed as the relatively new 'media archaeological approach':

> [F]irst, the study of the cyclically recurring elements and motives underlying and guiding the development of media culture. Second, the 'excavation' of the ways in which these discursive traditions and formulations have been 'imprinted' on specific media machines and systems in different historical contexts, contributing to their identity in terms of socially and ideologically specific webs of signification. (Huhtamo 1994)

These guidelines enable me to investigate the sense of déjà vu that is so often present in ontological accounts of media history, without falling into the trap of implying that we are 'simply re-running that which has come before' (Park, Jankowski & Jones 2011: xi). Indeed, media archaeology, with its explicit roots in the work of Michel Foucault, shuns the idea of seeing history as producing a continuous series of clearly defined media identities, focusing on the specific discursive formations of heterogeneous media manifestations instead.[11] Huhtamo as well contends that recently media scholars have begun to realise that a Foucaultian '(re)placing' of histories of media technologies 'into their cultural and discursive contexts' can offer us a far more complete understanding of history as 'a multi-layered construct, a dynamic system of relationships', in which the function of discursive regularities in media development is recognised (Huhtamo 1994).

Taking a media archaeological approach will also allow me to step away from unduly myopic approaches to the study of media development and use an interdisciplinary research method instead. As Huhtamo and Jussi Parikka note in their edited volume on the emerging media archaeological field, '[s]uch "nomadicism," rather than being a hindrance, may in fact match its goals and working methods, allowing it to roam across the landscape of the humanities and social sciences and occasionally leap into the arts' (Huhtamo & Parikka 2011: 3). The main advantage of using an interdisciplinary approach is thus that I will not be constrained to rigid paradigmatic borders that prevent the excavation of fruitful cross-references. In what follows I will therefore present hermeneutic analyses of discursive regularities in myths of the communication sublime, conducted from the perspectives of media history (Basalla 1988; Winston 1998) and comparative media studies (Boddy 2004), while drawing upon relevant theory and knowledge from a diverse field of disciplinary frameworks. These frameworks range from sociology (Desroche 1979; Castells et al. 2007), mythology (Campbell 1949), and philosophy (Bloch 1959) to evolutionary psychology (Dennett 1995; Gould 1997), literary history (Wegner 2002), and communication theory (Peters 1999). By taking on a multi-sided perspective that adheres to the guidelines of a media-archaeological approach, I intend to uncover the regulative function of utopian narratives and their various manifestations in media history.

Finally, by operating in this manner, I aim to extend the existing body of research on mobile communication to include a more expanded (and, in some respects, a more *holistic*) view of the development, marketing, and reception of mobile communication technologies. Ever since mobile telephones started to become truly commonplace in everyday life in the mid-1990s, there has been a growing scholarly interest in their nature and function, but most of the studies conducted to date are sociologically and empirically driven, and seek to describe and explain very specific demographically dependent uses and effects. Although such investigations can result in highly useful insights concerning mobile telephones and adoption strategies (Ling & Helmersen 2000), fashion statements (Katz & Sugiyama 2006), inter-family negotiations over ownership (Davie, Panting & Charlton 2004), expressions of subcultural identities (Skog 2002), changes in social interaction (Campbell & Russo 2003), and so forth, in general they accept the presence and proliferation of mobile communication devices as a given, and do not question the deeper underlying structures that have brought about the current mobile communication condition. The goal of this book is to address precisely this issue, and to provide a more comprehensive account of how mobile communication technologies have come to occupy such a prominent position in our media landscape. This does not mean that I will shy away from tapping into the vast amount of empirical studies available on mobile communication when needed, but large parts of my analysis will go beyond the micro-scale, and oscillate between a macro- and meso-scale.

Structure

The general structure of this book is determined by the four themes that have been identified in the recurrent mobile communication advertisement mantra: 1) a utopian belief in progress, that 2) engenders and reinforces ideas of perfected communication, which 3) find their expression in media evolution that 4) has bred the current feverish mobile communication condition. My analyses of these four themes are presented in two parts. In part I, 'Venturing into the familiar unknown', a theoretical and historical framework is constructed on discourses of hope, purpose, utopia, and progress (chapter 1) and on quests for improvement found in communication theories and models (chapter 2). In part II, 'Where angels speak', I use this framework to uncover myths of improved communication in media evolution (chapter 3) and reveal the practices and paradoxes of idealised ideas of communication in the discourses of mobile communication technology (chapter 4). In the following, a more detailed overview is presented.

In chapter 1, I start out by analysing the lineages of idealised ideas of communication, which I track in recurring discourses of utopia and progress. In order to ask the question whether in the evolution of communication media we can distinguish a recurrent urge to reach the communication sublime, I start by investigating the notion that the reachability of any sublime, whether it is religious, technological, or communicational, presupposes a fundamental need to gauge the status of a contemporary situation, and then to project a vision of an ultimate better place onto another time. Given that there are linear, cyclical, and spiral models of time, I describe how each of these models give shape to the idea of a better place, and then, in a brief historical overview, pinpoint several moments in time when the urge for a better future, expressed in stories of hope and purpose, has created visions of utopia and progress. Because reaching the sublime is by definition impossible, because utopia and progress always interlock with dystopian and regressive views, and because technology has also been the cause of many disasters, I conclude this chapter by exposing the contradictory and paradoxical elements that can be found in projects that aim to arrive at the sublime.

In chapter 2 an inquiry will be made into the various conceptual models of communication found in communication theories, particularly in regard to each theory's angle on the concept of 'ideal' or 'perfect' communication. I first focus on the spectrum of myths surrounding the communication sublime, and on how these myths, while presenting paths to utopian worlds by offering a heterogeneous collection of solutions to the problems of misunderstanding, remain situated in largely theoretical or fantastical narratives. This spectrum of idealised ideas of communication is structured into three main categories, which are formed by the paradigms of one-to-one communication, one-to-many communication, and many-to-many communication. From there I shift my attention to how these idealised ideas of communication, when we follow them to their radical vanishing points, invariably seem to lead to propositions of a world view that, in reality, is untenable, and ultimately paradoxically negates any need for actual communication. Subsequently, this observation calls for an analysis of how myths of idealised communication are strategically used and deployed by social actors in order to have them serve specific goals.

In the second part of this book, chapter 3 examines how idealised ideas of communication interlock with the evolution of communication technologies by looking at new media discourses from a historical and comparative perspective. First, I give an overview of the main positions in the debate on which theoretical approach best covers the paths that the development and use of technology traverse. My main focus here is, as in the two previous chapters, on the role and persistence of progress-minded models and on their sceptical counterparts. Arguing that we need a twofold scheme to study the continuously shifting position of technology in society, one that accounts for differences in what seems similar as well as for similarities in what seems different, I then proceed by describing an evolutionary approach to studying technology development. This approach aims to account for the multitude of factors that make media evolution an erratic and unpredictable process, as well as recognise the myths, hopes, and beliefs that have always been present and influential during technological development. This is then put to the test by using the theoretical framework established in part I to construct an encompassed understanding of how myths of the communication sublime manifest themselves in media evolution. I give a historical overview of media development prior to (and partly overlapping) that of mobile wireless technology, in which I trace continuously recurrent ideas of idealised communication placed against a backdrop of social, cultural, political, and economic factors that also played their part in this process.

In chapter 4, I carry over the perspective on the evolution of communication technologies from chapter 3, and continue by specifically focusing on historical accounts of how mobile communication devices have been developed and how they have been able to pervade everyday communication situations. Again, I aim to investigate how these accounts are rife with utopian thought and ideas of progress, and as such could be seen as a continuation of the discourses of potential found in the imagining and marketing of earlier media. Next, I shift to an evaluation of how such myths of the communication sublime in discourses of mobile communication technologies are both seemingly fulfilled and at the same time confronted by the paradoxical consequences of that fulfilment. Through an analysis of the four most characteristic promises in the discourses of mobile communication, which I see as those of ubiquitous connectivity, fluid sociability, real-time relief of anxiety, and omniscience through ever-present knowledge, I intend to find if these do indeed largely meet the basic criteria required from a system of pure communication; yet, I also ask whether they increasingly foreground the inherently terrifying nature of such a system. I then end this chapter by making the claim that the search for further improvements in communication will continue, and I elaborate on this by briefly tracing the paths that idealised ideas of communication take in the present day towards discourses of location-based services and of the 'Internet of Things'.

This book concludes with the observation that although there is no constitutive deterministic or teleological force in the evolution of communication technology per se, there is, however, a strong tendency to attribute utopian and progress characteristics to new media, prompted and inspired by a human need for myths that tell of reaching the communication sublime. These myths have made many appearances throughout time, and many technological projects have reacted differently to them, but their presence has been constant. It is the dominant discursive elements of current mobile communication technology's discourse that represent the hallmark of the myths of the communication sublime, which is to potentially contact anyone, anywhere, at any time. But, just as with older media, mobile wireless technology will never allow us to completely excise the problems of communication.

Part I

Venturing into the Familiar Unknown

1. Discourses of progress and utopia

> *As is so often the case in history, what a seer, prophet, or philosopher believes to be the result of novel and original inspiration almost invariably turns out to be a remembered insight from some perhaps long-forgotten book or author.* (Robert A. Nisbet 1980: 117)

In order to construct a theoretical framework for a macro-scale perspective on idealised ideas of communication in the ongoing development of media, it will first be necessary to examine the nature and functioning of hopeful expectations, utopian myths, and beliefs in progress. This will not result in an exhaustive overview of the body of work done on utopian thought, as a vast amount of literature has already been written on this topic.[1] The historical account that I construct here will serve to describe the various ways in which longings for a sublime state have expressed themselves through time, so that we are able to gain knowledge of how the concepts of 'utopia' and 'progress' acquired and changed their meanings, and how they have subsequently been incorporated in communication theories and models (the subject of chapter 2), in accounts of improvement in media evolution (the subject of chapter 3), and ultimately in mobile communication media discourses of the present day (the subject of chapter 4).

This chapter will assess the underlying premise of the existence of ideas of progress and utopia, which I contend is ultimately rooted in a human urge to hope and search for purpose and meaning. This urge expresses itself in 'necessary fictions', regulative narratives that foster a sense of improvement by positing two situations that differentiate in time and then presenting the latter of those two situations to be in some way better than the former. Evidently, the ways in which these necessary fictions function are very much connected to how relevant cultural concepts of time and the future are structured, so a short overview of three basic perspectives of time (linear, cyclical, and spiral time) will be given. These perspectives of time are then traced throughout cultural and social history, exposing their influence on ideas of progress and pinpointing the moment where utopian thinking arises as a particular subset of these ideas. We will see that the perennially recurrent hopes and beliefs in necessary fictions foster a poignant and paradoxical conviction that humankind can actually pursue a progressive path towards utopia. Through an analysis of dystopian and anti-utopian world views, which always serve as shadows to utopian projects, I will show that future-oriented necessary fictions invariably and tragically interconnect with notions of cleansing, repair, and, ultimately, reunification with a sublime state. In the end, the ultimate paradox is that humans cannot stop to 'name the unnameable final destination, to construe the unconstruable question about the meaning of human existence' (Zipes 1988: xxvii).

Hope, belief, and purpose

> *Erwartung, Hoffnung, Intention auf noch ungewordene Möglichkeit: das ist nicht nur ein Grundzug des menschlichen Bewußtseins, sondern, konkret berichtigt und erfaßt, eine Grundbestimmung innerhalb der objektiven Wirklichkeit insgesamt.*[2]
>
> *(Ernst Bloch 1959: 5)*

Hoping and believing in the existence of purpose are arguably part of the most powerful and significant emotions in the lives of human beings. Without a general idea of where we are heading to, what is asked of us in the meantime, and above all some indication that things are going well, life becomes hard to endure. Moreover, in most times of hardship a basic drive towards self-preservation starts to manifest itself, which is geared towards just one thing: to improve conditions. Making life endurable can mean ensuring that there is enough bread, water, or electricity to get through to the month's end, but it can also take on a metaphysical form, as many world religions, new age communities, cults, self-help groups, philosophical endeavours and psychological sessions continue to show. It is through the cognitive acts of hoping and believing that ideas of purpose are constructed and sustained: while a strong belief may preclude the need for hope, and hoping for something does not necessarily entail that you believe it will happen, both hoping and believing can be employed to orient oneself, in the present, to a specific instance of the future, which, more often than not, is projected to be a better one.

The value of hoping and believing in making meaningful everyday experiences as well as extreme events has widely been recognised by a diverse range of academic fields. In sociology (Desroche 1979), theology (Moltmann 1967), psychology (Snyder 2000), and neurobiology (Newberg, D'Aquili & Rause 2002; Newberg & Waldman 2006), hope and belief have been described as fundamental in shaping a person's sense of purpose and well-being in life. What many studies share is their stress, first, on hoping and believing as relying upon a system of affective forces that provides both shelter (from fear or suffering) and direction (towards a better place), and second, on the premise that this system is accessible to anyone. Particularly notable in this respect is the work of Austrian psychiatrist Victor Frankl, who put forward one of the most compelling accounts of the benefits of having something, anything, to believe in. As a former Auschwitz concentration camp prisoner, he wrote about his experiences in *...trotzdem Ja zum Leben sagen: Ein Psychologe erlebt das Konzentrationslager [Man's search for meaning]*, in which he described how prisoners who believed that there was still some purpose or meaning left during all the suffering, and oriented themselves towards a future event or task, had less trouble surviving. For Frankl, this undertaking was more than just a mental trick; he refused to reduce the search for meaning to a mere '"secondary rationalization" of instinctual drives' (Frankl 1992: 99). Hence, as psychology scholar Paul Wong has noted, Frankl places the source and continuous presence of the need to hope and believe squarely in the human spirit, which houses 'love, *the will to meaning, purpose,* creativity, conscience, the capacity for choice, responsibility, sense of humor, etc.' (Wong 2001, emphasis added).

What makes Frankl's contributions valuable for the understanding of utopian beliefs in general and of ideas of progress in communication technologies in particular, is that he posits a future-oriented frame of mind as a given. His notions of hope, belief, and purpose reflect a conviction that they constitute essential properties of what it means to be human, and that we cannot (or should not) do otherwise than employ them. Moreover, as a therapist with a keen interest in behaviourism, Frankl saw the search for meaning guided by an innate hierarchy of values, which he thought was 'founded on our biological past and [...] rooted in our biological depth' (Frankl 1992: 146). In other words, he held that the drive to look for purpose and meaning is a constitutive feature of humans, and that it will therefore necessarily express itself in our actions, cultures, and, last but not least, in (grand) narratives of how things came to pass, with what purpose, and with what direction.

As such, Frankl's views are in line with those of the school of evolutionary psychology, in which biological traits are seen as constitutive elements of the ways that humans reason. While reductionist explanations for certain types of human behaviour can in some cases be quite controversial,[3] they do provide exemplary material to support the contention that to hope for and to believe in a better life are undertakings that are shared across time and cultures. Evolutionary biologist Daniel C. Dennett, for instance, maintains that human organisms come equipped with a cognitive skill called a 'hyperactive agent detection device', which has, over thousands of years of evolution, instilled in us 'the urge to treat things [...] as agents with beliefs and desires' (Dennett 2006: 117). As a result, this urge led to the notion of deities, 'agents who had access to all the strategic information' that humans did not possess (ibid.: 126). Thus, the need to believe in God or Allah or any other supernatural, omnipotent, and omnipresent force that guides and steers us is, according to Dennett, the outcome of a biological and evolutionary process, one that favours the human organism's susceptibility to having the comforting thought that 'something' or 'someone' has made sure that we exist and live for a reason. In other words, Dennett holds that this mode of making understandable an otherwise chaotic world was favoured by natural selection as a fruitful means to survive, thereby preserving and extending the strand of genes that is carried by us god-fearing survival machines.[4]

As will be elaborated further in chapter 3, such evolutionary models of why we attribute purpose to what we do in life provide a useful analytical framework for understanding the mechanisms underlying the development of media. But, I should add, they are not without pitfalls. The notion that meaning and purpose are part of our biological make-up, and even might constitute the driving force of evolution itself, is not undisputed. The other side of the debate on the biological nature of purpose is reflected well by the work of evolutionary biologist Stephen J. Gould, who refused to adhere to the almighty power of the gene in gradually steering evolution towards a determined destination. According to Gould, the evolution of biological entities progresses without set goals; there is no purpose to be found in its directive principle (Gould 2002). Moreover, in response to Dennett's attempt to describe the presence of just such a directive principle in *Darwin's dangerous idea* (1995), Gould held that sociobiological explanations of human evolution and behaviour are 'just-so' stories, which only create the false illusion of a consistent, contingent, and goal-oriented system, whereas empirical research has so far not delivered substantial evidence to sustain the actual existence of such a biological blueprint (Gould 1997).

While Gould raises a valid point about the problems of the ease with which a need for empirical evidence is discarded, his strategy of dismissing sociobiological arguments by stating that they merely offer a compelling story does not address the questions of why there is a human search for purpose to begin with, or what is the purpose of hoping and believing in ideas of progress. And very important questions they are: considering Victor Frankl's findings, notions of purpose and meaning are what drive us – perhaps, thus, up to the point where they lead us to perpetuate compelling and deterministic stories that tell of their embedment in evolution. What remains to be explained, then, is the fact that the logic of purposiveness continues to have a strong foothold, as exemplified by stories of ever-present improvement in communication technologies. The act of searching for purposeful meaning and then finding it within directive systems may not necessarily be the result of an evolutionary effect, but it is undoubtedly a very powerful and recurrent trait.

The answer may lie in how the human psyche works, if we follow culture critic Steven Shaviro; purposiveness is actually not something we discover, he holds, but *attribute*, because we need this 'necessary fiction' in order to make sense of our surroundings (Shaviro 2003: 209). Thus, while we are very well able to acknowledge that we cannot understand a world without purpose, we should not mistake its presence for proof that it is a *constitutive* element in the process of evolution; instead, it performs a *regulative* function. This view on the need for purpose as a regulative function that produces necessary fictions coalesces with what Vincent Mosco calls 'a remarkable, almost willful, historical amnesia' (Mosco 2004: 117), meaning that without continuously and deliberately forgetting that earlier attempts to reach an ultimate goal have failed, we will not be able to find purpose in trying again, nor persist in (re)telling myths that disclose the ultimate route.[5] This, as I will argue later in this chapter, is the prime reason for the fact that so many utopian projects resemble each other: advocates of these projects simply cannot escape the lure of the ever-present potential, while knowingly ignoring the fact that similar projects have never before been fully actualised. They become both victims and proficient users of stories that tell that there is purpose, and because there is purpose, there is direction. This is the regulative function of necessary fictions.

Necessary fictions: Expressing the need for explanations

Religion is the greatest utopia to have appeared in history.
(Antonio Gramsci n.d. cited in Desroche 1979: 80)

The argument constructed so far is thus that, although there is no need to postulate a purpose in evolution, we need it so much to rearrange and order the perceived chaos in our daily experiences of life that we incorporate it in necessary fictions. Vincent Mosco's engaging account of the persistence and omnipresence of myths also suggests that these necessary fictions constitute an intrinsically human condition, found everywhere around the world and transcending local cultural traditions. As such, an analysis of myths could very well explain the feverish notions of unlimited potential found in discourses of mobile communication technologies, something that will be developed more fully in chapter 4. But I am getting ahead of myself; in order to understand how necessary fictions lead to utopian ideas of progress and subsequently to a need to improve communication technologies, it will first be necessary to investigate the main reasons that make myths appear particularly in progress-minded narratives, and to what extent the acts of hoping and believing orient us towards directions that seemingly lead to a fulfilment of utopian dreams.

One influential explanation for the recurrent nature of myths of progress is that searching for, and hoping to find, the ultimate reason of why things happen as they do is a fundamental part of human storytelling tradition. Comparative mythologist Joseph Campbell has contributed abundantly to the view that human history is driven by this search for universal truths, continuously reinventing and retelling stories that narrate of ultimate journeys to those truths. Moreover, in *The hero with a thousand faces* (1949) Campbell argued that these stories all follow a single narrative pattern, that of the *monomyth*, of which thousands of variations have been derived. Borrowing the term from a passage in James Joyce's *Finnegans wake* and partly basing his understanding of it on Jungian archetypes, Campbell constructed an elaborate model of the monomyth, which in short and adapted by Fredric Rice reads as:

> The hero is introduced in his [sic][6] ordinary world, where he receives the call to adventure. He is reluctant at first but is encouraged by the wise old man or woman to cross the first threshold, where he encounters tests and helpers. He reaches the innermost cave, where he endures the supreme ordeal. He seizes the sword or the treasure and is pursued on the road back to his world. He is resurrected and transformed by his experience. He returns to his ordinary world with a treasure, boon, or elixir to benefit his world. (Rice 2001)

Note here the 'innermost cave', which lures the hero into starting his quest; how the hero is 'resurrected and transformed', indicating he is renewed, has progressed, has changed for the better; and how he returns with 'a treasure [...] to benefit his world', affirming that the risk of the journey was all worth it. The story structure is very familiar, and contains plot elements we easily take for granted. Using any derivation of this monomyth it is possible, as Roland Barthes already noted, to give the elements that make up the myth a 'natural and eternal justification, [...] a clarity which is not that of an explanation but that of a statement of fact' (Barthes 1972: 143).

The power of the monomyth is thus that it can be used as a general template for countless ideologically loaded stories, all of which will contain a readily recognisable and accessible structure that, through retelling, is reaffirmed again and again as the blueprint for how basic facts of life manifest themselves. A better place really is just around the corner, it says; if only we believe it is, and take risks in getting there, we will be rewarded and everyone will benefit. As already exemplified in the introduction of this book, advertisements for mobile communication technologies make full use of this type of necessary fiction. They present us with an almost transcendental journey towards a place where everyone can connect to each other, reach mutual understanding with ease, and will have left a world of frustration and obstacles behind. As such, they uncannily reflect the basic narrative structures of many religious stories of transcendence, which tell of reaching a religious sublime state like heaven, nirvana, or satori.[7] According to Campbell, such similarities between religious themes and imaginative stories are no coincidence; he found support for his argument that the monomyth is the single most powerful narrative structure we know in the fact that he could recognise it in the dogmas and utopian outlooks of all world religions, on which he wrote extensively in his series of books *The masks of God* (1959-1968). He held that there is a single, unknowable truth underlying all religions, an immanent force that has given life to everything and into which everything will return in the end, one that lies behind the creation of many if not all of the world's stories.[8] The power of this force is immense; the mere fact that at the present day more than 85 percent of the world's population declare themselves as religious,[9] shows how many people in one way or another are confronted with the question of purpose, and try to approach that question in a meaningful way by immersing themselves in the narrative myths that the religious sublime supplies.[10] A necessary fiction 'purpose' may be, but it is a very potent one that can claim cultural and religious independency.

Another way to approach the manners in which utopian thought and ideas of progress find their continuous and instantly familiar expressions in narratives is, according to Campbell, through a probing of our collective unconscious, which both sustains and is sustained by religious scripts, folk stories, fairy tales, urban legends, and the like. Whether it is produced through mystical a-priori knowledge or through common experiences, shared instinct and shared culture, our collective unconscious holds dreams that reflect myths, and produces dreams that reinforce myths. The functioning of these dreams has to be taken quite literally; as literary critic Leslie Fiedler noted, myths are experienced as 'projections of certain unconscious impulses otherwise confessed only in our dreams, but which once raised to the level of full consciousness serve as grids of perception through which we screen our so-called "reality"' (Fiedler 1996: 34). And indeed, just as Campbell's analysis of religions uncovered a drive towards a single shared truth, studies in psychoanalytical research have shown that there are many similarities found in dream patterns across a wide variety of cultures (Domhoff 1996; 2005), suggesting that, again, there is a shared unconscious understanding of expressions of what Campbell calls the universal or transcendent truth. This understanding subsequently informs our actions in our waking hours, directing us along the path of the monomyth.

In the evaluation of the nature of necessary fictions that tell of purpose, utopian progress, and final resolutions, we have to be careful, however, not to settle, as Campbell seemed to do, for the existence of archetypes as their prime driving force. Similar to the problematic issues of controversial arguments in evolutionary psychology, the reduction of human behaviour and storytelling to mere effects of archetypal determinism in the unconscious can lead to a rigid conception of imagination, and does not adequately explain how visions of a better future are translated into a myriad of different actual actions. Here it is instructive to turn to the work of Ernst Bloch, German philosopher of revolutionary utopianism and author of the seminal multi-volume *Das Prinzip Hoffnung* [*The principle of hope*] (1959). Like Campbell (albeit within a broader political framework), Bloch studied folk stories, fairy tales, and myths in order to find shared characteristics of visions of a better life. Unlike Campbell, however, he did not think that looking for Jungian archetypes was a productive exercise. He was highly critical of Jung's reworking of the Freudian unconscious into a collection of 'primeval memories or primeval fantasies' (Bloch 1959: 62), and found that this prehistoric vision of the unconscious – one that harboured archetypes as primary drives – was dangerous and objectionable, as it fed reactionary and irrationalist tendencies that could lead to a justification for fascism.[11] Moreover, it left no room for imaginative heterogeneity; used as templates, archetypes work as constraints, suggesting and almost dictating that derivations are impossible. Instead of seeing the unconscious as a static and rigid entity, Bloch wanted to explore how a radical and revolutionary new could be instigated from it, and how notions of such a process were envisioned and expressed in stories of utopias, hopes, and beliefs.

In order to make room for a possible analysis of the radically new, Bloch introduced the concept of the not-yet-conscious, which he posited as a supplement to Freud's unconscious. We do not really know ourselves, according to Bloch, so we cannot extract everything there is to know about ourselves from our unconscious.[12] The not-yet-conscious, then, as scholar of folklore and fairy tales Jack Zipes writes in his introduction to Bloch's selected essays, is 'formed by the impulse of hope, in which inklings of what [humans] might become manifest themselves. For the individual, the not-yet-conscious is the psychical representation of what has not-yet-become in our time and its world' (Zipes 1988: xxxii). The not-yet-become is thus the domain of everything-that-could-be in our daily reality, and the not-yet-conscious is formed by stories of hope that tell of how the not-yet-become, at some point, might come to be. The most important thing about these stories, Bloch insists, is that they need to be productive, they need to uncover themselves at spots where they can form an impetus for action. The places where Bloch traces these manifestations of hope are primarily in *day*-dreams, and not in dreams, as Jung and Campbell argued. There is a significant difference between dreams and daydreams. Whereas the former are part of the unconscious and 'house repressed and forgotten desires and experiences', writes Zipes, daydreams fall into the realm of the semi-consciousness and therefore 'can be productive for the formation of individuals and the world since they [...] point to real, objective possibilities' (ibid.). Daydreams can inform possible ways to take action, and give shape to the means by which humankind expresses what these roads to reaching certain objectives might look like. This is why, as Zipes summarises Bloch's viewpoint, 'all art and literature that have anything to say to humankind are utopian' (ibid.). Utopia is not just a repressed ideal that keeps churning out different versions of the same monomyth, it is an actual, objective goal that can be reached by acting upon engaging and hopeful daydreams.

Bloch's conception of the not-yet-conscious thus not only adds another dimension to how we can understand the recurrent nature of utopian myths that tell of improvement and ultimately of reaching a sublime state, it also stresses that these myths can be identified as having continuous and real effects on the creation of a future-oriented mindset – even more so than when approached from the perspective of the monomyth or archetypes. They *do* something with us, they make us susceptible to the lure of that-which-lies-before-us, and as such they make us conscious of the fact that every time they are told they can set us in motion. They constitute, in several ways of interpretation and differentiated in time, an insatiable part of human reasoning that is aimed at bringing about change for the better.

This is why myths work so well in utopian stories, and what makes their presence so persistent throughout time. Grand ideas of where we are heading to, or what we should do to improve things, grab hold of our imagination and propel us into action, but each time they do so, the paradoxical nature of the ensuing quests makes us think that we are not quite there yet, and the whole process starts again. The recurrent tone of hope is ever present; as sociologist Henri Desroche concurs in *Sociologie de l'espérance [The sociology of hope]*, necessary fictions of hope, belief, and purpose pass through recurring historical cycles, and lead us to realise that '[n]o route has ever led any caravan to reach its mirage; *but* only the mirages have set the caravan in motion' (Desroche 1979: 144, emphasis in original). Desroche is quick to add that the process of hoping for and constructing plans to realise a better life is not inevitably cyclical; neither is it 'necessarily *linear* as the optimism of a cumulative progress rising to infinity would postulate', but it 'could be a *spiral*', connecting both the recurrent linear and cyclical aspects of the 'panorama of millenarian phenomena' he finds in human history (ibid.: 42-43, emphasis in original). Pondering the thought that the '"messianico-religio-ideologico-revolutionary" whole', the concept with which he describes the matrix of articulations of the hope for a better world, 'constitutes a prophetic dimension of collective consciousness' (ibid.: 143), Desroche makes clear, as I have also argued so far, that 'to orient oneself to a better future' is an inherent and shared human property, but, it has to be noted, one whose description and articulation are dependent on a vision of time as linear, cyclical, or spiral.

Concepts of time and progress

> *[T]he question to be addressed to [the utopian desire] is, does that desire work to pull the present forward, progressively, towards the as-yet-inexpressible but hopefully better future state, located in what Ernst Bloch calls the future unconscious, or is it a regressive impulse, in search of some prelapsarian lost domain located in the* past *unconscious.*
>
> *(Jan Relf 1993: 108, emphasis in original)*

Meaning or purpose can be obtained by envisioning a more tolerable future place that, when crisis is over, lies in front of us. In all guises, this thought that somewhere, just within reach, lies a better world, is arguably as old as the moment humans began to make sense of their surroundings. Describing this better world, setting out paths, undertaking the supposedly necessary or needed steps to reach it and actually getting there is what is often called progress, a phenomenon that must be approached quite carefully. Throughout history and within different cultures, ideas of progress have manifested, and continue to manifest, in many forms, depending on what concepts of time and future those cultures hold. To construct a schematic overview of how ideas of progress evolved and exerted influence on the creation and sustenance of myths of communication improvement, it will prove to be useful to group the many existing varieties of visions of time and the future into the three basic categories that Desroche proposed in his book on hope. Sohail Inayatullah, co-editor of the *Journal of Future Studies*, describes these categories as three basic shapes: 'the linear evolutionary shape of progress (the dominant paradigm of development), the cyclical shape of the life-cycle and the natural world, and the spiral shape that combines progress and tradition' (Inayatullah 2005).

Clearly, according to Inayatullah, it is the linear shape in particular that is associated with progress, as it lends itself quite well to creating a mindset based on making predictions. What lies further ahead may not be directly accessible, but at least this 'further' holds the inherent possibility that there is a direct path or a general direction that can be taken in order to arrive at a desired point in the future. Progressive linearity denies degeneration, evades the idea of going back, and tries to hold an equilibrium while continuously elevating it. There are, of course, various periods of short-term ups and downs that become visible when one zooms in on a small scale, but ultimately, and on a larger scale, the net result projected by the linear perspective is advancement. Cyclical time, on the other hand, knows only movement leading ultimately back to its origin. There are linear aspects to be found in cyclical time as processes can temporarily go 'up' or 'down', but the process eventually ends at the beginning. In cyclical time, writes Inayatullah, 'change is normal and opposites exist in dynamic tension in every stage' (ibid.). There is no general direction – in the linear sense – towards which cyclical time is heading, there are only recurrent phases.

There is reason to doubt, however, that the concept of progress can only be found exclusively in linear perspectives. While linear views are not seldom seen as optimistic ('a better future lies just ahead') and cyclical views as pessimistic ('we will inevitably return to from whence we came'), both schools recognise there are positive as well as negative characteristics of their perspectives to be pointed out. Linear time harbours promise and prediction, but also causes catastrophic scenarios when wrongly taken paths that lead to 'progress traps' cannot be abandoned (Wright 2004: 5). Cyclical time preserves traditions and holds no unpleasant surprises, but can create a feeling of powerlessness as the fate of history cannot be escaped. As a result, if the idea of progress is to be squarely coupled to a model containing at least some positive connotation, both linear and cyclical visions of time provide ample room for its incorporation. The fact remains, however, that both visions are slightly handicapped metaphors; they do not fully capture the idea that progress also knows its moments of déjà vu.

The ready answer to this problem is provided by the spiral vision of time, which combines linear and cyclical time in order to connect progress with history. According to Inayatullah, the spiral pattern intends to remove the future from the confines of pre-determined history, from the cycle, and to create the possibility for [...] an acceptance of structure, but a willingness to transform the suffering associated with history, and to find previous pockets of darkness and illuminate them, to pierce through silences. (Inayatullah 2005)

As such, the spiral view recognises recurrent processes, concepts, and ideas, as well as an imminently present higher plane towards which these processes, concepts, and ideas can lead. Thus, progress in spiral time can take on the incremental and accumulative properties of linear time, as well as the traditional and value-preserving properties of cyclical time, without having to yield to a single, restrictive framework.

The spiral, then, owing to its intricate integration of both linear and cyclical attributes, is a powerful tool to express causal relationships and to provide explanations of events for which there exists an urge to find their purpose. In the spiral vision of time, events take place because the linear course of movement leads us in a certain direction, or because the events happened before and must happen again. Capturing the spiral in Ernst Bloch's terms, hope works on the premise of already having sensed but not yet experienced what we are looking for; utopia has already existed, in ourselves, now we need to return to ourselves to find it again. The reason for the spiral's seduction therefore lies in its self-referentiality, a property that is of significant importance for understanding ideas of purpose and progress, and which will be described in more detail later in this chapter. Because the spiral connects itself to itself, it constructs, by definition, a constellation of everything that can possibly happen, continuously moving and incorporating new events, interpreting them in terms of old ones.

At this point it must be stressed that none of these time visions imply that causality between events is a given; rather, *perceived* causal relationships in the experience of time are more often than not induced by a psychical need that, quite similar to the functioning of necessary fictions, asks for an imposition of order on chaos. This urge for a 'spatialisation of time', as Henri Bergson (1910) called it, underlies the distinctive teleological elements of determinism so often found in ideas of progress, and masks the open and indeterminate character of time. Spatial configurations of time quite easily lead to a depiction of time as a coherent and seamless flow, whereas in an open-ended view of time, just as the outcomes of successive throws of dice are not influenced by each other, each unit of time can be seen as qualitatively distinct from every other. Bearing in mind this distinction between diachronic time as a research object and synchronic time as a framework for a historical approach, the next two sections provide an overview of how cyclical, linear, and spiral time each are seen to have had their dominant periods in history, co-shaping ideas of progress and utopian stories along the way. This overview should be seen as a necessarily selective look at a heterogeneous body of ideas consisting of what can be called 'historical constellations of utopian thought' (Manuel & Manuel 1979: 15ff); not all stories share the exact same structure or themes, but they do enlighten the multitude of ways in which the future-oriented mindset has expressed itself through time.

Visions of time and progress until the Renaissance

> *We dream much of paradise, or rather of a number of successive paradises, but each of them is, long before we die, a paradise lost, in which we should feel ourselves lost too.*
> (Marcel Proust 1924: 186)

While our current day-to-day experience of time can give rise to a fairly strong feeling that the structure of life is governed by causes and effects, and therefore always already has allowed for a linear realisation of goals that have been set in the past or present, this is generally thought by history scholars to be a fairly recent and mostly Western vision of progress. Lacking the experience of a swiftly changing world, the argument goes, people for a long time did not see any apparent direction that history was moving towards, and this attitude was more or less carried over until at least the Renaissance.[13] A majority of studies of ideas of progress therefore hold that only when the field of science established discrete systems, in which accurate measurements delivered undisputable proof of advancement and thus of the accumulation of knowledge, only then could the linear idea of progress as we now understand it develop and claim its important place in Western civilisation (Ginsberg 1953: 7).

Although there is much merit to this view, if only because the exact wording of what we now mean by 'progress' cannot be transposed to other historical periods such as classical antiquity,[14] I would suggest that myths that tell that humankind can advance to higher levels of civilisation and will continue to do so in the future are not unique to the last few hundred years, but have had a regulative role in many eras, in various shapes and roles. The modern era did not invent the idea of progress, but intensified and politicised its mythical underpinnings to an unprecedented extent, so much so that we are still entangled in it to this very day, and experience its influence on discourses of media evolution. The present section will substantiate this claim by showing that the human inclination to maintain a future-oriented outlook based on hope and a belief in purpose can be traced in many more and earlier times. The observations presented here should not be interpreted solely as constructing an account of chronological developments, but more as presenting a number of systematic and representative cuts in a historical timeline, and thus as showing the broad genealogical transformations in the thinking of time, progress, and utopia.

In most present-day historical research it is generally acknowledged that accounts of the structure of time in ancient – mostly Greek – sources emphasise its cyclical nature. Time cycles were believed to follow an eternal arc that stretched from a Golden Age to a Silver, Bronze, and finally, an Iron Age, after which divine intervention reset time back to a Golden Age. In these 'ages of the world', systematically written down for the first time around 700 BC by Hesiod in his Ἔργα και Ἡμέραι [*Works and days*], humankind typically experiences a slow but sure moral and spiritual decay, from a peaceful and happy time during the Golden Age to a dismal and chaotic epoch ruled by war and greed in the Iron Age.[15] In this cyclical chain of events there is no progressive improvement to speak of, only a progressive degeneration from a once sublime state that only at an indefinite point in time could be restored. Most early Greek poets and philosophers therefore rarely looked ahead, instead turning their attention to the Eden-like past and emphasising the deplorable state of their own society. Technological developments may have provided the most palpable examples of how the Greeks might have experienced something that resembles our current notion of progress, but these developments were more often seen as bringing decline instead of advancement.

By describing man's transformations in life as part of a cyclical and recurrent journey away from an ideal place, and positing them as ubiquitous and fixed truths about the human condition, the logic that connects the ages of the world is made up of two influential forces of human hope, and thus of desire and action: first, a feeling of loss caused by separation from one's origin, and second, the recognition that there will never be a final stable state, that things are bound to happen again and again. Both forces engender their own specific necessary fictions. The first conveyer of desire and action, a feeling of loss, is in Greek mythology most prominently visible in Plato's *Symposium* (written some four hundred years after Hesiod), in which he recounts how the poet Aristophanes tells of the creation of the two sexes. This creation myth describes how, long ago, humans came in three genders (androgynous, male, and female) and lived in perfect harmony. But, because the humans became complacent and did not properly honour the gods – especially Eros – they were punished and split in two, resulting in a division of humankind in heterosexual men and women, homosexual men, and lesbians. Since then, every one of us is eternally searching for our lost half, hoping to be reunited again in a state of bliss, under the guidance of Eros. The story outline thus acutely sets up an erotic longing for a return to a sublime state, propagating countless quests aimed at reaching that state; as Eros does not exonerate us, however, it is a longing that can never be fulfilled. Nonetheless, the story stirs us into action because it presents us with a goal that lies in the future: there is a path, and it should be travelled.

The second influence upon desire and action, the acknowledgement that things never remain as they are, gives expression to feelings of nostalgia as well as of hope; both look in opposite directions for a point in time where things were or will be better. According to historian Robert Nisbet (1980) this opposition can be observed in any discursive account on ideas of progress, and therefore it is possible to argue that not all Greek philosophers and poets saw their time on Earth as part of a cycle of regress. Moreover, even though Hesiod is often associated with having a regressive view on history,[16] he could actually be seen as writing about progress in a very familiar modern fashion. In a second infamous myth in *Works and days*, Nisbet exemplifies, Hesiod tells of Prometheus, who stood up to the will of Zeus and single-handedly helped humankind make its first advance to civilisation by stealing fire from Mount Olympus and giving it to the pitiful humans. Then, having 'lifted mankind from its primal degradation and misery to a level where man might seek to rival the gods', Prometheus was punished by Zeus for all eternity (ibid.: 19). Thus, from that propitious moment on, humans were endowed with reason and gained the almost inescapable drive to build upon self-made stepping stones and progress to a better life. While Nisbet admits that it is not exactly clear what the nature of this progress was – Hesiod did not differentiate between moral, ethical, technological, or other forms of progress – it is a vision of progress nonetheless, and it had a considerable impact on classical thought. As Nisbet continues to show, throughout the classical period, Hesiod's suggestion of the possibility of an unlimited advancement of humankind has influenced many writers, such as Aeschylus, Plato, Aristotle, Lucretius, and Seneca, all of whom had respect for 'knowledge [...] that gives protection, comfort and well-being to mankind', had a clear conception of 'the acquisition of this knowledge by man through his own abilities', and recognised that this acquisition had 'taken place cumulatively over a period of time' and would continue to do so (ibid.: 46).

We therefore see that, while the cyclical model of time is held by many historians to be dominant in myths of life in classical times, complementary notions of linear progress can also be distinguished, thus negating the view that those myths told of nothing but recurrent decline and degradation. This observation has a significant impact on the current project of tracing the genealogy of stories of improvement. The Greek (and later Roman) society is widely held to be the starting point of Western civilisation, of the origin of many seeds of present-day thinking, so if ideas of improvement can be found amidst a *Zeitgeist* that is supposedly free of any belief in linear advancement, it makes their relation to contemporary ideas all the more potent. This is of course not to say that ideas of progress have not changed for more than two thousand years, or knew no alternatives. Although I am inclined to disagree with historian Morris Ginsberg when he says that 'we are not justified in ascribing to [progress] universality, continuity or necessity' (Ginsberg 1953: 5), because in my view there *is* a certain widespread psychological necessity to entertain a – any – notion of 'that which is better', I agree that any attempt to reduce the myriad forms of ideas of progress to a single unbroken strand is futile. Rather, all these variants should be taken into account, composing a broad and contextually sensitive perspective on the manifestations of ideas of progress. The conditions that influence how we look at our own contemporary existence are dynamically structured, constantly shifting the boundaries of what is considered as 'better'.

Linear ideas of progress found in myths in classical times were thus overshadowed by cyclical thinking, and differ from the contemporary version we have come to know. Not only was the progressive accumulation of knowledge more an individual than a collective process (one that did not necessarily entail social or ethical advantages), classical ideas of linear progress are also generally not endowed with much potency: according to historian Sydney Pollard, they were more 'poetic explorations' than 'firmly based philosophic views', and served no systematic search for a universal law of progress (Pollard 1968: 2-3). Accounts vary as to the moment in time when allusions to the modern notion of progress did find their first articulation; while most historians follow the argument that a belief in progress 'could not take a firm hold over men's minds' until the doctrine of Providence was abandoned when modern science emerged (Ginsberg 1953: 7), others squarely place the seeds of the modern idea of progress much earlier, in Saint Augustine's *De Civitate Dei* [*City of God*]. In this chief orthodox Christian work, Nisbet, for example, already sees notions of 'a unity of all mankind', 'historical necessity', and 'a confidence in the future that would become steadily greater and also more *this*-worldly in orientation as compared with *next*-worldly' (Nisbet 1980: 47).

Whether the work of Augustine indeed presents us with the first glimpses of the roots of modern progressive ideas, or, as Pollard asserts, was actually 'written to *combat* the doctrine of mundane progress' (Pollard 1968: 5), what it did do was introduce to Christianity – and consequently to much of Western philosophy of history in the ensuing centuries – the originally Jewish notion of millenarianism, the 'expectation that the end of the world is near and that, accordingly, a new earthly paradise is at hand' (Noble 1997: 23). This apocalyptical view of the future of humankind marked an important change in the dominant understanding of time – from cyclical to finite, and thus linear – as well as initiated a goal-oriented pragmatism by foretelling that a paradisiacal state would await the chosen ones. The basic elements that make up the ingredients of the monomyth can be seen to express themselves: the universal truth lies at the end of a long road, and people have to travel this road in a certain way in order to actually arrive at that truth.

The growing conviction in Christian thinking that a new millennium was at hand, however, is not necessarily seen in historical studies as the fundamental driving force behind a change in the general attitude towards the manipulability of the fate of humankind. Many historians discussed so far believed that, long engulfed in a mindset of inferiority, Christian thinkers in Europe did not assume it was possible or even appropriate to actively pursue the worldly implications of the Final Truth. For orthodox Christianity, progress along the earthly road meant spiritual progress only, obtained by abiding by religious dogma and passively waiting for God himself to announce the Final Judgement. It is Robert Nisbet, again, who disagrees with the dominant vision of the Middle Ages as a period in which 'thought was lost in contemplation of the heavenly hereafter and in despair of or disdain for the things of this world'; he names Joachim de Fiore as the most powerful of prophets, whose ideas of a future earthly paradise would make their way from the 'twelfth down to the nineteenth' century and inspired numerous thinkers of progress (ibid.: 93-95).[17] When we look at the multitude of myths present in the latter of these centuries, there are strong hints that, after the Renaissance, the resigned attitude towards the inevitability of fate slowly morphed into a growing belief that man was capable of altering his living conditions for his own benefit after all (Gombrich 1974).

The birth of utopia and its progressive realisation

> *A map of the world that does not include Utopia is not worth even glancing at, for it leaves out the one country at which Humanity is always landing. And when Humanity lands there, it looks out, and, seeing a better country, sets sail. Progress is the realisation of Utopias.* (Oscar Wilde 1891: 16-17)

In his book *Utopie en kritisch denken* [*Utopia and critical thinking*], Dutch philosopher Martin Plattel notes that although the urge to transcend everyday reality and strive for a better life has seen many different historical manifestations from classical antiquity and onwards, it was not until Thomas More wrote his 1516 novel *Utopia* that stories of human endeavours towards progress began to gain true momentum in the socio-cultural imagination (Plattel 1970: 27-30). Key in this respect was that More gave a name and a place, or, rather, a 'non-place' (from the Greek οὐ τὸποσ [ou topos]) to this envisioned better world: he described Utopia as an island where all inhabitants lived together in harmony, thereby making this a 'good place' (derived from εὐ τὸποσ [eu topos]) as well. Although More portrayed Utopia as a fantasy,[18] and did not consider it to reflect his own personal political and social ideals, he cleverly imitated the mixture of fact and fiction that was characteristic of stories of discovery in his time (such as those of Marco Polo), thereby making *Utopia* believable and successful, and essentially founding the genre of the utopian novel.

As is predominantly the case with imaginative projections of possible ideals on future or not-yet-existing worlds, *Utopia* was born out of dissatisfaction with the contemporary state of affairs. Elaborating on Plato's ideas on the ideal configuration of government of the city state in Πολιτεία [*The republic*],[19] More presented an alternative societal structure to that of Britain, France, and other European feudalist nation-states, indirectly criticising their unwillingness to create a more humane system and put into practice important ideals of Christianity. The characteristic that can be recognised as most significant about Utopia is that More did not describe it as a natural, untouched paradise to be found, but as a society that could be constructed. In other words, *Utopia* is a narrative that meets – or essentially constitutes – Ernst Bloch's basic prerequisites for expressing the not-yet-conscious, the daydreams that should be acted upon in order to bring forth truly utopian societal change.

Utopian fictions such as *Utopia* thus present us with an imaginative other place that can be interpreted as the ultimate reward for quests for a better life. As narrative theorist Phillip E. Wegner points out in *Imaginary communities* (2002), utopias are '"nowhere" [...] precisely to the degree that they make *somewhere* possible, offering a mechanism by which people will invent anew the communities as well as the places they inhabit' (Wegner 2002: xvi-xvii, emphasis in original). This theme of perceiving the world as open to human discoveries and hospitable to utopian ventures can clearly be seen as gaining significance in the sixteenth and seventeenth centuries. Written at a time when Christopher Columbus and Amerigo Vespucci had just made their great expeditions to the West, *Utopia* very much exudes New World pheromones, mingled with an invitation to re-invoke Saint Augustine's image of the City of God, a perfect place on Earth. In the 1620s and 1630s, around the time Francis Bacon wrote his *Utopia*-inspired *The new Atlantis*, the utopian theme was literally exported from the British homeland by the Puritans. Their main goal was to complete the English Reformation and establish a truly spiritual society; by shaping the foundations for this exemplary righteous community, a New Jerusalem, they created a utopian backdrop against which many later American communities were modelled.[20] As anthropologist Steve Mizrach notes, America became 'the vehicle for Europe's utopian imagination, [...] a place where the regeneration of the age promised by the Rosicrucians and other groups might come about, home to bold experiments in the investigation of nature and society' (Mizrach 2001).

Social experiments such as those of the Puritans, an increasing stream of publications that aimed to describe all-encompassing models of how the world worked,[21] a rise in the number of New World expeditions: these are all reflections of an intensifying tendency to act upon utopian daydreams of progress in the seventeenth century, an era that is marked by a 'growing confidence of the power of man over his environment and, ultimately, over his destiny' (Pollard 1968: 13). Thomas More was of course not the sole instigator of this change, nor was he merely the nomenclator of an idea that 'had always already existed in some natural, ideal realm' (Wegner 2002: 27), but *Utopia* became the reference work for anyone thinking about social reform in the following centuries. In the wake of the Reformation, which had already signalled a slow change in the dominant attitude towards the fate of inescapable degeneration and the circularity of time, the theme of societal manipulability gradually took hold, expressing itself in necessary fictions while surfing on the rising tide of a new idea of progress, which was soon to find its zenith in the Enlightenment.

The Age of Enlightenment, which roughly spans from the second half of the seventeenth century to the end of the eighteenth century, is widely held to be the period during which a rational belief in linear progress both truly emerged and reached its highest peak. At no other time do we find so much optimism about the future and man's role in constructing a utopian version of it than in these centuries. It was then that the belief in progress acquired its definition as '[meaning] that mankind has advanced in the past – from some aboriginal condition of primitiveness, barbarism, or even nullity – is now advancing, and will continue to advance through the foreseeable future' (Nisbet 1980: 4-5). It is also the time when a linear sense of history manifested itself more fully. The prime force behind this belief in linear progression was reason; human rationality became seen as the most powerful way to break free from irrational fears, dogmatic beliefs, and dangerous ignorance, and to establish standardised methods for engendering improvements.

The fields in which these improvements were best visible were the mechanical arts and sciences. This was not only due to the refinement of instruments, but also, and more importantly, because the accumulation of scientific knowledge could seemingly go only one way, and that was forward through generation upon generation. The Apocalypse still had a prominent place in Enlightenment views of the world's fate, but humankind no longer had to suffer while waiting for the subsequent Paradise. Moreover, progress in the mechanical arts and sciences became seen both as a *sign* of immanence of the awaiting Paradise and as the *cause* of this immanence. A certain historical need for continuous improvement thus arose: without it, there would be no progress, and without progress, there would be no redemption (see Tuveson 1949: 153-203). As will be further elaborated in chapter 3, this is one of the main themes running through myths of technological evolution from the eighteenth century onwards: the mechanical arts became commonly viewed as the driving force in history that brought a human-made earthly paradise closer with each new invention and improvement, and this view still resonates in present-day conceptions of technology.[22]

What is clear to discern from the events that took place in this period of changing worldviews is that there was an intimate correlation between the production and distribution of necessary fictions that told of reaching a better place through processes of progress. Together with the heightened significance of the gathering of knowledge, as well as the conviction that improvements in knowledge would automatically result in an improved society, the amount of utopian texts published during the Enlightenment rose to an unprecedented number.[23] While such an outburst of creativity could have signalled a wide diversification of themes, the original blueprint of More's *Utopia* always remains visible. The contents of

Enlightenment utopian texts invariably focus on the distribution of wealth and labour, the configuration of the governing political system, and on true justice and morality; the solutions to the problems posed by these questions all deal with specifying strict relations between groups and individuals, defining socially determined spatial zones, and placing the utopian city radically outside of contemporary societies yet in reach of those that really wanted to change things (Baczko 1978: 30-38). As would become clear in later centuries, this enormous gulf of imaginary and projected social ideals set the tone for many of the plans for social revolutions in Europe, as well as for the birth of modern positivist thinking.

At this point it should be noted that, in the accounts of historical manifestations of ideas of progress and utopian thought up until and including the Enlightenment, only two of the three visions of time that were discussed earlier are seen to have expressed themselves most visibly, namely cyclical and linear time. However, although their narrative expressions have indeed known dominant periods, they did not possess any exclusivity. The cyclical worldviews in classical Greece were accompanied by linear subplots, and the linear stories of progress during the Enlightenment were often coupled to the desire to return to a reconstructed Eden, the Paradise before the fall of man. This also means that the third vision of time, the spiral vision, understood as a blend of cyclical and linear time, had always been potentially present, although it had not manifested itself explicitly in Western thought. It is therefore fitting that, at a time when both visions of cyclical and linear time had clashed violently in a major paradigm shift, just after the end of the eighteenth century, German philosopher Georg Wilhelm Friedrich Hegel put forward an idea of time and history that resembled a spiral-like movement towards the Absolute.

For many historians writing on ideas of progress, Hegel is a mandatory stopover. According to Nisbet, there is 'scarcely a work in Hegel's voluminous writings that is not in some fashion or degree built around the idea of becoming, of growth and progress' (Nisbet 1980: 276). Similarly, Pollard writes that Hegel's work represents the view that 'the actual development of the world', or what Hegel called *Geist* [*Spirit*], necessarily and unequivocally marches forward towards a state of complete consciousness of itself (Pollard 1968: 89). This progression is the result of a dialectical mechanism: every phenomenon ('thesis') is met by opposition ('antithesis'), out of which, on a new and higher level, another phenomenon ('synthesis') arises, and then the whole process starts again. As Charles Van Doren notes in his review of Hegel's position among authors who see progress as the result of some natural cosmic principle, there is thus a 'rhythmic pattern [...] to be observed in all phenomena', in which '[e]verything comes to fruition, then to grief, then to a higher truth' (Van Doren 1967: 91). The spiral emerges: the process by which *Geist* heaves itself up is repetitive and therefore circular, the direction towards the Absolute is linear.

Hegel's account of the structure and motion of time and history thus stood for unchecked progress towards a final goal, which strongly resonated with the utopian mindset of the time. Plattel even states that the structure of *every* utopian thinking follows the basic Hegelian triad: 'it moves away (antithesis) from the status quo (thesis), and focusses on a better future (synthesis)' (Plattel 1970: 47). What is important to stress here, though, is that humankind's role in Hegel's view was that of an intermediary: we have little to no agency in the forward movement, it is *Geist* that progresses and makes its strides visible through the actions of humankind. Moreover, according to Hegel, attempts to actively enforce utopian conceptions of progress form the necessary dialectical counterpart of universal will, and are destined to bring about Terror. Published at a time very much characterised by the French Revolution and its grim aftermath, Hegel's description of progress as the manifestation of a natural cosmic principle deeply contrasted with the deeds sparked by the utopian mindset of revolutionaries, or what he called 'specific individual acts of will' (Hegel 1967: 604). We will return to the dark and paradoxical sides of utopian projects soon, but here it suffices to say that, to Hegel, the image of the ideal Absolute Mind serves as the ultimate synthesis towards which the dialectical process itself moves, constantly setting humankind up for struggles between conflicting states of being.

Notwithstanding Hegel's reservations regarding the terrifying effects of specific individual acts of will, at a time when necessary fictions of hope and purpose became more and more occupied with the notion that an earthly paradise could be installed, Hegel's views on history and progress, as controversial as they were then and are now,[24] were, in a variety of interpretations, shared, used, or expanded on by some of his early nineteenth-century contemporaries (like Henri de Saint-Simon, Robert Owen, Charles Fourier, and, later, Auguste Comte) in their diverse utopian schemes. The theories they used and the objectives they pursued differed from person to person, but overall a shared conviction can be identified that progress towards a more just and more peaceful society was feasible through instigating a revolutionary change and then building society anew by using a combination of science-based knowledge and technologies. While the main criticism of these 'utopian socialists', as they were somewhat disdainfully called by Karl Marx and Friedrich Engels, was that they merely sketched fantastical visions that could never be realised, their writings have been very influential in shaping a more scientific view on the manipulability of society, and inspired the foundation of many utopian communities, again confirming the power of the utopian impulse (see Schehr 1997).

The period from the middle of the nineteenth century up to the present day is one that shows an oscillating pattern in the ways that ideas of progress and utopian thought were perceived. First, amidst accumulations of scientific knowledge and rapid technological developments that gave rise to the Industrial Revolution, in Western societies the modernist belief took hold that progress was not only feasible, but could be steered towards very particular notions of what a utopian future should look like. As Plattel concurs, the utopian mindset 'was no longer aimed at *finding* new possibilities, as it was during the Enlightenment, but at *making* and *creating* them' (Plattel 1970: 35, emphasis in original). New scientific discoveries and new technologies increasingly strengthened the idea that all kinds of futures were now wide open, waiting to be realised. The vision of utopia as predominantly separated from us by space largely made way for a vision that saw utopia more as distant in time, something that can be distinguished in modernity's necessary fictions, as reflected by the popular works and novels of Albert Robida, H.G. Wells and Jules Verne. From the beginning of the twentieth century onwards, however, ideas of progress lost much of their appeal due to large-scale failures of utopian social projects, and because there was a growing sense that new technologies could wreak havoc as much as they could propel humankind forwards towards a brighter and better future. Both world wars and other conflicts and catastrophes in the latter part of the twentieth century radically tempered the belief that humankind was indeed progressing towards a better life. Moreover, in what can be broadly described as a postmodern reaction to modernity's conceptions of historical progression, a highly discontinuous and contingent view of time and history arose, in which there was no room for postulating the existence of hierarchical or organising principles such as those attached to progress and utopia (Marx 1994).

However, despite intellectual and cultural mistrust of grand narratives that tell of a better life that should be within reach, elements of their discursive constructions remain visible in stories in popular culture, in prophetic books on the optimistic future of what has been called the Information Revolution since the 1990s and onwards, and of course in advertising and publicity material for the sophisticated problem-solving capabilities of networked communication technologies (Hughes 2004: 107-109). Especially industries in capitalist societies are very well aware that the idea of advancement is a necessary element to their existence. Therefore, while necessary fictions of hope and purpose have known many manifestations that were invariably allied to dominant visions of time, and thus cannot be perceived as having a single definitive function, they continue to create new expressions of utopian narratives and stories of progress, which in turn persist in influencing decisions, processes, and actions in everyday life. The lure of thinking in terms of absolutes, universals, or totalities when daydreaming of better futures may have diminished in the latter part of the twentieth century, but it has not been discarded completely: much like in the spiral vision of time, the experience of a rhythmic pattern of rise and decline is today still coupled with a belief in a more general forwards movement. This also means that the inherent paradoxes of such spiral-inspired utopian thinking remain nurtured in conceptions of better futures. As we will see, the paradox forms the tragic disposition of necessary fictions in general, and of communication myths in the mobile age in particular.

When utopia meets dystopia: Behold the paradox

> *To be sure, I believe that without the notion of an unfettered life [...] the idea of utopia, the idea of* the *utopia, cannot even be thought of at all [...] There is something profoundly contradictory in every utopia, namely, that it cannot be conceived at all without the elimination of death; this is inherent in the very thought.* (Theodor W. Adorno 1964 cited in Ziper 1988: 10, emphasis in original)

So far in this chapter it has been argued that necessary fictions are part of a psychological need to create order out of chaos, to create utopian landmarks that we can look out for while travelling along the paths of life. Because, inevitably, a sense of hope is projected upon the creation and preservation of such landmarks,[25] the genealogy of ideas of progress and utopian narratives has up to now predominantly been discussed in the light of how they present us with a pristine state of 'better'. What makes utopian landmarks so powerful as blueprints for hoping for the better is that, because they are foremost experienced as exactly just that, a blueprint, they can take on *any* interpretation of what exactly constitutes a better place. As a result, the utopian mindset becomes a locus where all possible interpretations come together to form a heterogeneous composite of deeply ingrained longings. Moreover, this imaginative constellation is so complex that, in the end, it cannot exist except as the paradoxical fulfilment of an unfulfillable desire. This paradox is inevitable: the human yearning for resolution, as a manifestation of the belief in a purpose to our existence that is recurrently expressed in narratives of progress and utopia, will always generate as well as face paradoxical problems when confronted with actual endeavours to achieve a final or other ultimate state of being. In a tragic yet unavoidable fashion, humankind cannot fully embrace this view; it vehemently tries to cope by either orienting itself towards a final change, putting its faith in transcendental ideals and envisioning the implementation of perfect societal structures, or by combining all these views in a belief in a millenarian apocalypse, the ultimate revolution that will establish a thousand-year-long divine heaven on earth before the end of time comes.[26]

This inherent paradoxical inclination is very much a recognised element in studies of the utopian mindset, and thus should be included in any analysis of how future-oriented processes – such as those that can be found in the mobile communication condition – try to deal with or even conceal conflicting aspects of their projected outcomes. Approaching the specificity of utopian thought from a psychoanalytical standpoint, English scholar Jan Relf adds that utopian fiction

> narrates and stages imaginary, fantastic solutions to an unassimilable contradiction in the human condition; it strives to reconcile the knowledge that we

inhabit an irretrievably fallen and divided world in which the ideal state is unattainable, with the irresistible and mysteriously present idea of unified perfection (the good object) which we continue to desire. (Relf 1993: 110)

Building upon Jacques Lacan's ideas about the manifestation of desire, Relf here points to the notion that utopias stand for a state of otherness that we long for continuously without ever attaining it, and that they thus inherently confront us with the question with what methodology they are ever going to be achieved.[27] The focus on paradoxes in utopias at this point is therefore prompted by the supposition that an analysis of their nature and of the struggles to resolve them will provide the means to further understand the discursive strategies that are employed by myths that tell of reaching the unreachable communication sublime. In order to conduct such an analysis, I will highlight three mutually nonexclusive ways in which paradoxes reveal themselves: first, as regulative principles for understanding life itself in general and its myths in particular; second, as solutions for blending unavoidable contradictory opposites in utopian myths; and third, as centres of unstable balancing acts that bring forward anti-utopian and dystopian views, which both complement and reaffirm the tantalising power of ideas of progress and utopian narratives.

Understanding life through paradoxes

The nature of paradoxes is such that they first and foremost inspire debates on conceptual notions of how we see the world. Often presenting more questions than answers in a continuous test of what is true, these debates traditionally centre around age-old philosophical puzzles that concern themselves with investigations of purpose and meaning. Instructive in this respect is the work of Roy Sorensen, who in *A brief history of the paradox* writes that paradoxes are the 'atoms of philosophy because they constitute the basic points of departure for disciplined speculation' (Sorensen 2003: xi). When proposing to use an analysis of paradoxes as a means to describe the history and focal points of philosophy, Sorenson holds that these '[b]odies of conflicting evidence' have fascinated us from the moment we began to ask ourselves where we come from and where we are heading to, thus stressing their prominent role in necessary fictions (ibid.: 1, 4). Most important to note here is that in every example of paradoxical thinking that Sorenson subsequently explores in his book, he identifies characteristics of paradoxes that are fundamental to an understanding of both our existence and its purpose-giving mechanisms, and therefore should be seen as elemental aspects of necessary fictions that aim to explain the inexplicable.[28] In other words, every utopian myth will contain these characteristics, and by describing any of them, the paradoxes in future-oriented blueprints can be uncovered.

On the level of making sense of our existence through telling necessary fictions, paradoxes are thus inevitable. A very compelling argument for this human inclination towards paradox can be traced in existentialist theories, especially in Martin Heidegger's treatment of *Gelassenheit* ('releasement') and Jean-Paul Sartre's description of its conceptual opposite *mauvaise foi* ('bad faith'). Whereas *Gelassenheit* deals with fully accepting one's existence as something that has no intrinsic goal or pre-given content, as something that can only receive its significance through the meaning one chooses to give to it, *mauvaise foi* is the result of *not* accepting the open-ended nature of our existence, of continuously asking for reasons and trying to find the answers outside of one's own will. Clearly, such a denial of things-as-they-are and of things-as-they-happen actively feeds and sustains a dual-pole system, in which paradoxes reside: there can be no coincidence when everything happens for a reason, and there can be no sense in it all when everything is contingent. People who live in bad faith – and there is 'a very great number' of them, according to Sartre (1956: 90) – often face and cannot accept the most obvious paradox: sometimes things are just what they are, even when they are not.

All this shows the double role of the paradox: it presents itself as a necessary element in understanding and coping with the mysteries of life ('Why are we here? What is our purpose?'), and it arises as the only way in which problems stemming from the desire to unite antinomies can be resolved, becoming visible in fantasies, fictions, myths, and daydreams of wholeness and completeness. In this sense, the paradox thus *defines* the utopian genre: the ultimate collection of definitive solutions to worldly conflicts can only exist, as Philip E. Wegner concurs, 'as a formal rather than a concrete spatial possibility, a figural space holder for something else that [is] finally unrepresentable' (Wegner 2002: 200). Wegner provides a well-crafted analysis of the relationship between utopian literature and the way it has both commented on and shaped the ways in which we view our world, and he notes as well that the utopian narrative typically functions as a cognitive space in which real world crises are finally dealt with. This classification alone could suggest that the main goal of studying utopian myths is to understand them as merely fantastic contemporary projections of a possible ideal future; however, like Wegner, I find that analyses of utopian texts can disclose them as 'rhetorical machine[s]' that 'participate in a significant way in the making of their social and cultural realities' while they strategically incorporate paradoxical constructions to highlight (and dissolve) opposites in those realities (ibid.: 37, 40).

Dissolving opposites in utopian myths

This brings us to the second way in which the paradox manifests itself in utopian narratives, which is in the ability to seemingly dissolve opposites into a homogenous whole. The emphasis here lies on 'seemingly', because the search for homogeneity is, again, a common structural theme in utopian myths that lays bare an inherent inevitability of creating unsolvable problems at the same time as they are solved. Universal answers invariably lead to new questions. As philosophers Hans Crombag and Frank Van Dun show in their book *De utopische verleiding [The utopian seduction]* (1997),[29] there are four basic and interdependent causes of conflicts that all utopian stories try to provide an answer to, and all of those answers prove problematic in finding an overarching logic that is designed to be valid for everyone. The four causes of conflicts that Crombag and Van Dun name are plurality, diversity, scarcity, and unrestricted access; the solutions are found in their opposites by providing unity, consensus, abundance, and 'righteousness'[30] (Crombag & Van Dun 1997: 27-28). While the authors see the establishment of a righteous society as the least imaginative of the four solutions – as this would merely need 'an external force of control' (ibid.: 28) – they find that the biggest problem lies in the idea that a well-functioning and all-encompassing consensus can be reached. Pointing to the human tendency to judge, value, decide, and act from one's own limited worldview, they note that 'even the most servile followers and rabid conformists are no match for the devastating powers of the cumulative misunderstandings, multifarious interpretations and wild guesses that make up the bulk of human communication' (ibid.: 30).

Solving scarcity by creating an abundance of goods is equally problematic according to Crombag and Van Dun; they note that to make sure there is enough for everyone, there either has to be a drastic restriction imposed upon all desires, yearnings, and needs, or that a great increase in productive power has to be achieved (ibid.: 32). The contrast is enormous: while in the first solution people find happiness in a return to a serene paradisiacal state where nature itself is a horn of plenty (a utopia that Crombag and Van Dun call 'ascetic'), in the second solution man conquers nature and literally manufactures his own happiness by producing everything he desires (what Crombag and Van Dun call a 'utopia of freedom') (ibid.: 32-33). Both solutions, however, ask for a wholly unrealistic organisational structure in order to either contain or satisfy all of a society's desires, and both fail to see that what is one person's serenity or freedom is another person's nightmare or enslavement.

In other words, there can be no utopia without what Bronislaw Baczko calls '*une représentation totalisante et disruptive de l'alterité sociale*' (Baczko 1978: 30, emphasis in original). This necessity of a regulative but paradoxical totalitarian praxis in order for utopias to work is identified by Wegner as well. In his analysis of Thomas More's description of the social and spatial organisation of Utopia, Wegner sees a continuous need for 'abstraction and homogenisation' in the way people dress and build identical cities in which everyone speaks the same language, follows the same customs, and abides by the same laws (Wegner 2002: 48). When showing that this application of universalising mechanisms automatically leads to the problems arising out of opposites described earlier, Wegner adds that 'universalizing tendencies themselves produce antagonisms [...] that threaten to tear the social structure to pieces'; the only way in which these unstable relationships can be countered is through the creation of 'an imagined set of particular allegiances and common grounds', what he calls an imaginary unity (ibid.: 48-49). In chapter 2, where I will analyse universalising patterns in idealised ideas of communication, we will again see how an imaginary unity (as embodied by the communication sublime) engenders processes in which, the moment that levelling solutions are used to combat certain problems, other problems are raised.

Moving away from utopia... to return again

Finally, we come to the third way in which paradoxes manifest themselves, which is through dystopian and anti-utopian narratives.[31] The observation that the inherent paradoxical nature of utopian narratives causes the creation of problems while simultaneously trying to solve them is one of the most prominent reasons that there have been many critical responses to utopian projects throughout time. As has been noted earlier, especially after the rise and fall of utopian socialism, with its tendency to propagate hypothetical visions of perfect societies without offering concrete and substantial mechanisms to create and sustain them, a contemptuous attitude towards the utopian mindset began to become commonplace, in which responses eventually took the form of outright condemnations of anything utopian as being unrealistic.[32] So, although scepticism is found in every era and there are earlier examples of literary works satirising the incessant search for utopia (Jonathan Swift's *Gulliver's travels* comes to mind), dystopian and anti-utopian narratives are typically found at the end of the nineteenth and the beginning of the twentieth centuries. Because dystopian and anti-utopian texts are written as a counterpoint to utopian narrative, and therefore define the utopian mindset by presenting themselves as logical opposites, it is important to focus on how these texts relate to the paradoxical character of utopia.

Dystopian and anti-utopian narratives typically exploit one or more fatal flaws in utopian designs, often by introducing an extrapolation of the supposedly ideal preconditions for utopias to exist and then extending them to their radical conclusions. As has been pointed out before, utopian flaws have to do with how universalising organisational schemas such as homogenising work, play, and housing, tend to leave out or stigmatise the particular. Subsequently, there is the problem of how the utopian hierarchy is to be established and maintained. If one thing has been made evident from the analysis of utopian thought, it is that an ideal societal structure will not be found in the wild, and therefore has to be imposed on an alleged non-ideal system.[33] This inevitably raises questions on who decides what is ideal, and who takes control to enforce the changes that are needed to reform society. As behaviourist B.F. Skinner, writer of the utopian novel *Walden two*, aptly remarked, '[d]esign implies control, and there are many reasons why we fear it' (Skinner 1981: 37). Common themes running through dystopian and anti-utopian narratives thus include a totalitarian control by an (unseen) elite or entity, the abolishment of identity, an outright loss of privacy, the complete submission to mechanised or otherwise technologised systems, and the inability to escape from a rigidly imposed social hierarchy. A brief analysis of two influential dystopian and anti-utopian narratives will serve to illuminate this point.

Probably the best-known dystopian novel today is George Orwell's ominous *Nineteen eighty-four*. In this work, we read that the lives of the citizens of Oceania are controlled by the Party and its omnipresent leader, Big Brother. Telescreens both broadcast Party opinion and monitor every citizen's activity, so that not only speech and behaviour but even thought can be kept within the prescribed Party regulations. Sexual desire is repressed, children are encouraged to spy on their parents, and cities lie in ruins as a result of never-ending wars between Oceania, Eurasia, and Eastasia. From these descriptions it becomes poignantly clear how Orwell thinks a totalitarian doom scenario can – or even *has* to – emerge when the necessity to keep everyone and everything in a supposedly (and subjective) ideal configuration is extrapolated towards its radical vanishing point.[34] Equally apparent is Orwell's focus on the paradoxical mechanisms that have to be employed by the Party in order to maintain its power. The Party is never wrong: in part because it continuously alters history,[35] but more importantly because it makes excessive use of 'doublethink', which is described by one of the Party's opponents as 'the power of holding two contradictory beliefs in one's mind simultaneously, and accepting both of them' (Orwell 1951: 220). Through the central role that Orwell gives to the Party's reliance on doublethink, exemplified most visibly in the Party's maxims 'War is Peace', 'Freedom is Slavery', and 'Ignorance is Strength', he shows how self-contradictory and impossible dogmas both automatically arise and have to be actively deployed when universalising projects fuse oppositional worldviews into a single tenable perspective.

A similar focus on totalitarian themes, struggles between collectivism and individualism, and finding ways to escape from it all is found in Yevgeny Zamyatin's lesser-known Мы [*We*] (1920), the book that can be credited as having influenced Orwell's *Nineteen eighty-four* and is seen as the twentieth century's first anti-utopian novel. In Zamyatin's fictional future world, 80 percent of the global population has died because of a two-hundred-year-long war that could be described as an archetypical clash between the modern technology-driven mindset and the prehistoric untouched wilderness of nature. Modernity prevailed, and the survivors gathered in a single city (the 'One State') that has shut out nature by means of a great 'Green Wall'. Zamyatin carefully employs common utopian mechanisms for the construction of an ideal society: everything in the One State is perfectly planned, balanced, and controlled through an intricate system of regulations, which, overseen by the 'great Benefactor', governs labour, free time, social and sexual relations, education, and justice. The 'Table of Hours' schedules and harmonises all activities of the city's inhabitants – who are not called citizens but 'numbers' – up to the precise second, and through the diary entries of protagonist D-503 the reader learns that (in contrast to Orwell's Oceania) this rational and mathematically structured life is considered good. But, and this is what makes *We* work as a counterpoint to classic utopian themes, underneath this narrative structure Zamyatin questions why the good life has to be paid for by removing every notion of individualism, creativity, and 'out-of-the-ordinary-ness'. Just as in More's *Utopia*, people in the One State dress the same, the city's architectural make-up is without any frivolity and the social hierarchy is strictly adhered to, but in *We*, D-503 slowly comes to realise that this is actually not perfection; rather, it works as a straitjacket for what he learns is his 'soul'. While at first he is of the opinion that the imposed rules of collectivism present him with true freedom because they relieve him from the burden of having to make difficult choices and take charge of his life himself, he gradually cannot help but notice a growing and deeply rooted desire to express his individuality. Diagnosed as having the 'soul-disease', he tries to resist the revelation that perfection is in fact unattainable, but he fails miserably. In the end, after having seen the chaotic and colourful world of the Mephi people behind the Green Wall, D-503's psychic state is in such turmoil that he decides he cannot leave his old world, and undergoes a 'fantasectomy' which cures him from having any sense of an autonomous self.

While this ending is of course a bleak one, one that enforces the assumption that *We* is essentially an anti-utopian novel because there seemingly is no resolution to the oppression of individuality in highly organised utopian societies, Zamyatin actually does present us with a possible way out of thinking of utopias as perfected but dreadful places. The answer lies not so much in persevering where D-503 had turned back, and choosing a Rousseauian return to the pure authenticity of nature, but more in an acceptance of contradictory worldviews and paradoxes as necessary structural elements of life. The irrational world of nature is no more utopian than the rational world of the city, but it needs to exist to establish a highly unstable but nevertheless persisting and necessary balance. Number I-330, the girl that made D-503 doubt the perfection of the One State in the first place and wants him to join her in overthrowing the status quo, makes this point clear when she tells D-503 there is always a next revolution: 'There is no final one; revolutions are infinite. The final one is for children: children are frightened by infinity, and it's important that children sleep peacefully at night...' (Zamyatin 1972: 174). Thus, Zamyatin tells us, as long as utopias are pictured as final, stable, and unchanging societal configurations, they will fall victim to their inherent flaws; the tricky solution to the problems of wanting to design an ultimate utopian place is to accept that the only possible 'ultimate' rests in embracing the presence of paradoxes. In Blochian terms, then, the dialectical infinite revolution described in *We* 'critically illuminates that which is "not yet" available to any of us' (Wegner 2002: 170), and as such serves as a useful – or even necessary – tool to help guide our desire to hope and imagine in a more productive way.[36] Seen as such, utopian thought regains some of the value it lost during the twentieth century, and comes to stand for a continuous receptiveness to change, albeit an absent-present one.

The utopian paradox thus comes full circle: on the one hand, we desire stable configurations that do not exist and cannot be reached (and even if we could realise utopia it would undermine the essence of what it means to be human); on the other hand, the escape from this desire in the form of an infinite revolution is in effect unimaginable itself, and holds within it the prospect of evolving *anyway* towards a different understanding of what it is to be human, leaving behind our present hopes and dreams of what constitutes a better future. Criticism of the flaws in utopian thought, when viewed from the perspective of the human tendency to look for purpose and meaning, quickly leads towards the equally utopian question 'If not envisioned like that, how *then* should we think of what constitutes an improvement upon what we have now?' Thus, dystopian and anti-utopian narratives may reinforce the notion that utopian thinking is something to be scorned, but in the end there is no escape from it; at least embracing the dialectical version, one that acknowledges the 'concerns of natural law and human dignity' as Wegner (ibid.: 171) notes, may provide us with outlooks that are less bleak than those often associated with radically developed utopias.

In the following chapter, the complex functioning and roles of these paradoxes in structuring and guiding the discourses of hope, purpose, progress, and utopia will be revisited as we look at how myths of improvement and fulfilment play their part in idealised ideas of communication. Problems similar to those described above arise when communication is thought of as something that lies at the heart of what needs to be improved in order to progress towards an ultimate and final better place. The price that is always paid for these kinds of converging solutions is a repression of antinomies, which are necessary to understand the longing desire per se. As Crombag and Van Dun note, realised utopia is a 'happiness machine' in which there can actually be no conscious knowledge of what happiness is, because its counterpart does not exist and therefore cannot help to construct a scale to judge the degree of happiness on (Crombag & Van Dun 1997: 49-51). The banishment of unhappiness and the accompanying search for a serene ideal may reflect psychological desires that are understandable, but what is wilfully forgotten is that the ultimate fulfilment of those desires will radically transform human existence, along with all notions of what constitutes happiness, individuality, society, or, as will become clear in the next chapter, communication.

2. Communication ideals, communication woes

> *What [the] eschatological belief in the 'information society' hides is the fact that, as the ideal of the universalism of values promoted by the great social utopias drifted into the corporate techno-utopia of globalization, the emancipatory dream of a project of world integration, characterized by the desire to abolish inequalities and injustices in the name of the imperative of social solidarity, was swept away by the cult of a project-less modernity that has submitted to a technological determinism in the guise of refounding the social bond. The ideology of limitless 'communication' – but without social actors – thus takes over from the older ideology of limitless progress.* (Armand Mattelart 2000: 120)

Echoing the central question that communication theorist John Durham Peters critically assesses in his book *Speaking into the air* (1999), what exactly do we think is so wrong with communication that it needs to be fixed all the time? Why is it that, while the general experience of communication in everyday life can reveal itself to be a complex but also a rewarding process of self-affirmation, knowledge gathering, or social bonding, there remains a poignant longing for an overall improvement of communication – one that underlies communication theories, managerial efficiency recommendations, marriage counsellors' advice, anecdotes of communication gone wrong, and, last but not least, advertisements for and predictions about mobile communication technologies? And what remains of our attitude towards this so-called communication problem if we follow the paths of utopian communication ideals to their radical conclusions and find that, as suggested at the end of the previous chapter, they paradoxically lead to a place where there is no longer a conception of that which we desired? Surely, there is a reason why ideas of progress and utopia recurrently pervade the thinking about communication?

This chapter aims to provide answers to these queries by turning its attention to the power of the communication sublime and its integral function in engendering discourses of progress and utopia in the evolution of communication technologies. In order to uncover the ways in which the myth of 'ideal communication' expresses its questionable discursive role as 'solver of all problems', the observations made on necessary fictions in the previous chapter will therefore be used as a framework for the analysis of various forms of idealised ideas of communication. We will see that, although there are many approaches to describing what communication is and in what configurations it can be understood as being improved, they all, explicitly or implicitly, relate to the idea of removing conflicts and misunderstanding from the process, and thereby set up a paradoxical relationship between the practice of communication and its projected ideals. As will be argued further in chapters 3 and 4, this strained correlation is recurrently visible in discourses of media evolution, and especially those of mobile communication technologies.

First, I describe where and how idealised ideas of communication reside in theoretical discourses and models of communication. In order to sift through and categorise the vast body of work in this field, I will step away from using a chronologically ordered overview, and instead draw from Peters' work and concentrate on a collection of specific and exemplary theories and conceptions of one-to-one, one-to-many, and many-to-many communication. As will become clear, what is considered to be ideal in idealised ideas of communication differs from model to model, but the common characteristic is the proposition that there is an ideal to begin with, and that this ideal can be reached by improving existing models of communication through the elimination of communicative noise (often by technological means) and thus by bridging the gap to the mind of the other.

Next, I question both the nature and the theoretical productiveness of the ideals found in the analysis of communication models. It will be demonstrated how visions of communication as a lossless transferral of information – the immediate and direct connection of minds – may seem to promise the banishment of miscommunication, but when these visions are taken to their extreme, they do not turn out as ideal as is supposed. We can even say that the ideal human act of communication *depends* on the possibility for miscommunication, that dreams of pure communication inevitably clash with the actual social experience of being with others. Nevertheless, these dreams play an important role in the discursive constructions surrounding every new medium, and therefore it is necessary to look at the motives behind the production and use of these dreamlike ideals. Placing media marketing strategies and industry decisions that use myths of improved communication within Rein de Wilde's (2000) models of future-making, this chapter ends by making clear that, again, there are many ideals of communication, and they are not merely reflections of imaginary desires in the human collective unconscious but are actively used, engaged, and re-enacted to serve very specific goals.

Tracing communication ideals

Models of communication are never neutral; each implies an ideal scenario.
(George Myerson 2001: 43)

Both following and extending Peters' classification of 'dialogue and dissemination' as 'two *Grundbegriffe* in communication theory' (1999: 35, emphasis in original), the communication theories and models discussed here are presented in three categories: one-to-one, one-to-many, and many-to-many communication. The theories and models selected for scrutiny here are in part those identified by Peters as examples of how we can understand both idealised and problematic conceptions of dialogue and dissemination. These examples are then supplemented with theories and models that reflect typical present-day thinking about the presumed effects and future of networked communication media. In each category, the following questions will provide the basis for investigating the theories and models involved: what problems in communication are identified; which ideal communication situations are therefore implied; with what means are those ideal situations to be achieved; and what problematic consequences automatically arise out of the use of those means, which then set up another quest for improvement?

One-to-one communication

Keeping it together – Socratic dialogue: The Greek philosopher Socrates is arguably one of the most important figures in the history of Western thought. His tutoring on eros, politics, and knowledge, available to us mostly through the writings of his most ardent student, Plato, still echo in many of today's debates on love, ethics, democracy, and education. Most interesting in this respect are his ideas about how all that knowledge should be 'birthed',[1] or communicated, to use an anachronistic term. For Socrates, this birthing process had to have an appropriate structure, one that he based on the ideal of dialectic inquiry through the use of dialogue. It is no wonder that Peters, in his foray to construct a comprehensive view on the nature of dialogue, chooses to approach Socrates as if the philosopher were a communication scholar *avant la lettre*: as Peters observes, the notion of dialogue as an ideal means of true communication is still present in current communication theories, giving dialogue 'something of a holy status' and making it 'the summit of human encounter, [...] superior to the one-way communiqués of mass media and mass culture' (Peters 1999: 33).

There are two connected and very intuitive reasons for dialogue, or face-to-face conversation, to serve as a blueprint for the ideal transferral of ideas, knowledge, and information. Firstly, and richly described by Plato,[2] Socrates saw the virtues of dialogue in its interactive and reciprocal nature. Every question can be answered with another question, every ambiguous utterance can be followed by a request to explain, every non-verbal sign can be picked up and acted upon. The presence of two equal individuals in the same space, questioning and debating without obstacles, is for Socrates a prerequisite for the true bonding of souls; there needs to be a close relationship between the communicating parties for them to actually connect.[3] This leads us to the second reason why dialogue can be seen as the preferred and ideal communication model: media such as writing physically separate the communicating parties and thus enlarge the danger of scattered messages and misunderstandings. Socrates was not so much opposed to the act of writing per se, but he lamented the distance it creates between the writer and the reader, which emphasises the impossibility to give feedback to or question the indiscriminately dispersed text. Dialogue is ideal in its staging of a physical and live encounter, and therefore – at least for Socrates – preferable as a communication model to ones where the conversing parties are separated in distance, and technological means need to be used to establish contact.

Dialogue is thus framed in contrast to mediated communication, or rather, dialogue comes to serve as the presumed 'natural' form of communication that is compromised when it is mediated. As such, its myths hold that one should strive for dialogue as the primary means to exchange knowledge. According to sociologist Michael Schudson (1978), this notion that we need a model of communication to which all communication processes should adhere, even if such a model is fictional, is implicitly present in our culture. Dialogue as 'conversational ideal', as he calls it, is 'not one concocted by social scientists [but rather] a widely shared ideal in contemporary American culture which social science has uncritically adopted' (Schudson 1978: 323), and it would be reasonable to add that this is not only so in American but in Western cultures in general. Although there are many arguments against promoting dialogue as the ultimate and universal form of communication – and we will encounter some of those arguments later on in this chapter – culturally grounded ideals such as those of face-to-face communication have had and continue to have a sizeable influence on the direction of media evolution. Moreover, they stand at the base of the classic and ongoing debate whether media are bad because they contaminate 'real' interaction, or good because they provide continuously improved simulated face-to-face conversation at a distance, bringing people together even when they are physically apart.

In short, the myth of dialogue tells us that 'true communication [is] personal, free, live, and interactive' (Peters 1999: 50). The implied ideal is that only in an environment where there is the opportunity for direct verbal and non-verbal contact is it possible to truly connect, and even when two people in conversation freely decide to agree to disagree, there has to be no misunderstanding of what both parties agreed upon. If any of these conditions is not met, for instance when dialogue becomes mediated, we are facing communication breakdown.[4] What this means for the use of words, themselves a medium, is dealt with in the following subsection.

Reducing semantic fog – Charles K. Ogden and Ivor A. Richards' *The meaning of meaning*: The idea that language is the necessary vehicle for our thoughts – if they are to be conveyed to the mind of the person we talk to – has been explored by many, but, as Peters puts forward, when it comes to entertaining the notion that the use of language should be streamlined, none have done so with more conviction than literary theorist Ivor Armstrong Richards and literary critic and writer Charles Kay Ogden (Peters 1999: 12-14). In the first quarter of the twentieth century, both Ogden and Richards realised that they were concurrently engulfed in attempts to understand the functions of language, and especially how misunderstandings in the use of language could be remedied. To tackle this challenge, they went on to write *The meaning of meaning*, in which they established a 'science of Symbolism' and devised a model describing the functioning of linguistic symbols (Ogden & Richards 1956: 8, 11). This model, known as the semantic triangle, consists of a diagram in which all the possible relationships between thought or reference, symbol, and referent are drawn.

The most important observation that Ogden and Richards made was that the relationship between symbol (or word) and referent (or thing) is an *imputed* relation; echoing the semiotic work of Charles Sanders Peirce, they held that only through conceptual thoughts in the mind can words refer to things. This immediately lays bare the crux of the communication problem Ogden and Richards wanted to address: too often, they noted, symbols are used as if their reference is inherently attached to a single referent, stemming from a 'once equally universal' belief that words 'meant' something by themselves (ibid.: 9-10). The problem was that the same words were used in multiple instances by a multitude of people with different perspectives, resulting in these words pointing to all kinds of referents and thus creating 'almost all the difficulties which thought encounters' (ibid.: 12). So, even if two people would have been able to construct the ideal conversational structure in the form of a face-to-face dialogue, they would still encounter misunderstandings because of differences in interpretation.

Here it should be plain to see how a fundamental human feature, that of being able to interpret language in multiple ways, is approached as if it were a problem, one that needs solving for the benefit of humankind. To Ogden and Richards, the use of words is a necessary but often crude and error-prone way of expressing thoughts, a process based on causal yet arbitrary and unstable rules, which, unless improved by employing a system of universal semantics, stand in the way of clear and unconfused communication. In this stance we recognise a concern over looming mental isolation: without a common descriptive framework of words and their meanings, thoughts cannot cross and human communication necessarily fails. Thus moving forward from their theoretical analysis, Ogden and

Richards proposed a solution to remove all this 'semantic fog'. The solution was to be found foremost in educating the public in the unambiguous use of words, with the underlying thought that familiarity with a 'Theory of Definition [...] not only conduces to ease of deportment in reasoning and argument, but offers a means of escape from the maze of verbal cross-classifications' (ibid.: 15, 247). In other words, learning to strictly adhere to a fixed set of sign definitions would solve most interpretation problems.

Expanding on this positivist solution in later years, Ogden adopted a pragmatic approach and went on to construct a new language system that inherently would suffer less from ambiguity. His 'Basic English'[5] was to be a universal language, a 'solution of the problem of Debabelization', consisting of only 850 words that had the effective expressive power of 5000 words and 'could do all the work of 20,000' (Ogden 1935: 7, 14).[6] Reducing the amount of words in a language might seem a slightly odd solution for the problem of misunderstanding, but according to Ogden this would force people to be exact in their wording and sentence building, and thus promote clearness. He found that there was still much merit in learning strict rules and grammar, and held that the 'idea at the back of the[se] old rules [is that] because our thought is based on language, *and because it is important for our thought to be clear*, a great respect for form might be a help in the development of our minds' (Ogden 1935: 1, emphasis added).

What can be recognised here is a yearning for clarity in the meeting of minds, which in Ogden's view is to be achieved by simplifying the semantic structures of language and so by facilitating progress through shared understanding. In this attempt, he joined a wide array of people who also searched for a universal language, worked on creating artificial ones, or claimed that their mother tongue was the only one that could rejoin all the people of the world in harmony. As semiotician Umberto Eco has shown in *The search for the perfect language* (1995), many such endeavours can be found in European history, all in one way or another invoking the tragic story of the Tower of Babel and interpreting it as telling of 'how a real wound had been inflicted on humanity, a wound that might, in some way, be healed once more' (Eco 1995: 17). Pointing out that the proposed healing processes are invariably aimed at finding or (re)creating the pre-Babelian 'one language, a perfect language, a language spoken by Adam with God and by his posterity' (ibid.: 34), Eco notes that this divine unity has been and continues to be a powerful source of inspiration when thinking about solving problems of coexistence (ibid.: 18-19). Therefore, and echoing the utopian impulse described in the previous chapter, we can uncover in Ogden's Basic English the underlying idea that through communication improvement a harmonious and unified grand togetherness can be brought about, where there is nothing but a blissful state of being.

To summarise, the myth of removing semantic fog holds that although people can freely engage in dialogue, their communication will still be hampered by differing interpretative frameworks unless these frameworks are perfectly synchronised. The success of Ogden and Richard's quest – a search for a 'utopia of a concourse of consciousness' as Peters (1999: 14) calls it – thus depends on an unambiguous and clear transfer of information. In the next subsection, we will see how this problematic desire for a clear transfer of information has, in the present day, come to dominate much more than just attempts to improve the functioning of language, and has taken on a more technological dimension.

Noise reduction – Claude E. Shannon's 'A Mathematical Theory of Communication': Notions of purifying communication, like those of Ogden and Richards, were popular amongst early twentieth-century thinkers,[7] and gained an extra boost in the middle of the twentieth century with an increasing demand for improvements in efficiency in communication technology systems. Exemplary for this quest for improvement was the work of mathematician Claude E. Shannon, who in 1948 wrote a paper on a new way to measure the dispersion of information – or, as he coined it, to measure 'information entropy'. His ideas, which were first published as 'A Mathematical Theory of Communication' in two issues of the *Bell System Technical Journal* and were popularised later in 1949 in book form with a foreword by fellow mathematician Warren Weaver, revolved around proving that it was possible to transmit digital data error-free across a channel, no matter how much noise interference there was in that channel.[8] As it turned out, Shannon's paper and subsequent book marked the start of a whole new field of scientific research later called information theory – first known as communication theory – geared predominantly towards the optimisation of technological data transfer and compression. Over the years, information theory gave birth to and became closely associated with many other applied informational disciplines, such as cybernetics and the study of artificial intelligence.

What is important to note here is that, originally, information theory and its related academic fields dealt with the quantity of data, not quality: having defined 'information' as a measure of uncertainty in received signals, Shannon was less concerned with the meaning of those signals than with finding ways to successfully receive them exactly as they were sent. In other words, it was technological perfection that Shannon was after.[9] However, quickly after its inception, information theory lost its exclusively mathematical and technological character and became something much more in popular and collective consciousness; by illustrating his proof with a schematic 'general communication system' (Shannon 1948: 380), Shannon inadvertently presented the world with a model of communication that did not decidedly stress its quantitative nature. Bearing a slight resemblance in its linearity to Harold D. Lasswell's model of communications,[10] which was completed shortly before Shannon wrote his Bell paper, Shannon's model effectively became a one-size-fits-all blueprint that seemed applicable to both quantitative *and* qualitative acts of communication, including the perfect transfer of thoughts and meaning.

As a consequence of this generalisation of the model's applicability, it was readily employed in a multitude of academic fields, engendering a significant cognitive shift. As Peters notes of this shift, things were increasingly seen as types of 'information' which were then cast in 'communicative structures': DNA became known as 'genetic information', the brain turned into an 'information processor', and people (even whole nations) could begin to truly understand each other if they 'shared information' and communicated better (Peters 1999: 23). This 'informationalisation of our perspective on the world', as Dutch philosopher Jos de Mul (2002: 129) calls it,[11] has increasingly led to the view that we can see things as bits of information in a communicative system like that of Shannon's model, moving unaltered and unaffected by noise from source to destination.

The ideals of communication that are visible in this model echo those found in the attempts of Ogden and Richards to minimise and ultimately eradicate semantic fog. Whereas in Ogden and Richards' case thoughts and meanings faced the danger of never coming across unaltered (because multiple interpretations of words themselves create semantic noise), Shannon's schematic diagram invites us to transpose that semantic problem into a technological (that is, a computational) process, in which the application of the right mathematical formula seems to promise to do the trick. When captured in Shannon's model, all that stands in the way of people, institutions, or nations connecting their minds and really understanding each other, it seems, is simply a question of perfecting the channel and filtering out the noise.

In the observations of these three theories and models relating to one-to-one communication, we can see the perceived problems reflected in a distrust of remoteness, a suspicion of language, and an aversion to noise; the underlying ideals and proposed solutions point to a common desire for closeness, unison, and a banishment of obstacles that stand in the way of error-free connecting. The solutions themselves, however, are all problematic in the sense that they constrain the process of one-to-one communication to that of an ideal type that has no real world counterpart, either because language can never be perfect or because too much expectation is invested in technology. These problems will be given more attention later in this chapter; now it is time to look at how idealised ideas of communication are expressed in situations where there is just one source but many receivers.

One-to-many communication

Sowing the seeds – Christian dissemination: There is a long history of examples of how those in power usually have the best means to make themselves heard, and how they make good use of mass communications to send their messages.

This is especially visible in the frameworks of religious belief systems. Chapter 1 already showed how key elements of the Christian tradition have had a significant impact on the philosophical and collective consciousness of Western societies, ranging from propagating the notion that the Garden of Eden represents humankind in all its sublime bliss, to the eventual adoption of a sense of progress that intended to (re)construct a paradisiacal state. The means with which such ideas were expressed and have been presented through time – often resurrected and appropriated, in different guises, to suit specific socio-political or economic goals – have been identified as the telling of necessary fictions. According to Peters, the rhetoric of the parable can be seen as the best representative example of how necessary fictions take on their shape in Christian sources, and the best way parables are told is by dissemination (Peters 1999: 51). It is in these dispersed religious narrative structures that myths are typically used to pass on certain values, which are derived from an interpretative reading of what constitutes the religious or transcendental highest good.

Dissemination, then, is the opposite of dialogue: instead of a one-to-one, interactive, and reciprocal exchange of minds, dissemination deals with the indiscriminate, one-way, and asymmetrical spreading of messages. It wants to do away with having to put a lot of dedicated effort into communicating with many people. There is not so much a distrust of dialogue to be observed in using dissemination for communicating a message, but a preference for not wanting to single out individuals as having more right than others to hear that message, and for creating the opportunity for many people to feel they can relate to it (ibid.: 53-54). Indeed, central to the Christian doctrine is the responsibility to treat everyone the same, no matter their background and deeds.[12] Illustrated best in the parable of the Good Samaritan, Christian neighbourly love extends to love of the stranger, even the enemy, and aims to include all people in kinship with each other and with God. For this reason, dissemination is an ideal way to communicate: it circumvents the problem of continuously having to talk to individual people by making sure that whole groups are reached at once, or, inversely, by making available a ubiquitously accessible message.

There is a catch to this approach, though. Even if dissemination might be the best way to ensure that no one is left out during the communication process, only those who know how to listen will receive the message as it was intended. This is especially so when speaking in parables, something the parable of the sower exemplifies.[13] Dissemination is thus predominantly receiver-oriented, and because the onus is on the receiver to decode the information she has acquired, this leaves all the more room for multiple interpretations. This does not necessarily mean that messages do not come across, because compared to dialogue more people will concurrently receive a message (which virtually guarantees that someone will understand what that message entails), and it might not even be imperative to derive a single meaning from what was communicated. And yet, dissemination needs to be aimed at reaching a large audience in order to secure a sizeable 'yield'. Moreover, when we take into account that religious doctrines are often conceived as tending to declare one truth, it should not be surprising to see that in notions of dissemination a tacit underlying assumption can be pointed out, namely that while it may be that only those who have ears to hear will hear the message, it ought to be just a matter of time and repetition of that message before everyone has developed the right ears to hear.

The idealised characteristics that the act of dissemination possesses, in short, are an inherent tendency to celebrate standardised communication approaches, and the ability to spread messages across a wide range of people, which makes sure that those messages will be heard and understood by more than just one person. These characteristics also make clear, however, that there is a fine line between communicating the one truth and hoping that some will hear it, and manipulating the masses. The consequences of making assumptions about the suspected effects of repeatedly disseminating the same messages, prominently visible in twentieth-century mass communication models, is discussed in the next subsection.

Mass models – propaganda and hypodermic needles: The power of dissemination, especially when coupled to large-scale communication technologies, was to find a disconcerting expression in the first half of the twentieth century. In the decades when more and more people came to own radio and television sets, from the Great War up to and extending after the Second World War, it became increasingly evident that media were crucial components in strategies to sway public opinion. In Germany, as part of National Socialist propaganda efforts, radio was monopolised by the state and put under the control of *Reichsminister* Joseph Goebbels. Successful German experiments with mechanical and electrical television were touted as the results of 'a great achievement of technology and the natural sciences', creating a 'national technological myth' (Elsner, Müller & Spangenberg 1994: 129). In England, multiple fake radio stations were set up to mislead and demoralise German soldiers and civilians (Rowen 2003), and in the US, President Franklin D. Roosevelt's memorable fireside chats created an intimate atmosphere in which he tried to reassure the American people and gain support for the involvement in the war. Idealised ideas of communication here take a far more economic form: as long as messages come across as intended and reach many people, their transfer has been successful.

These experiences of media use in wartime, along with the emergence of a powerful advertisement industry that also employed radio and television to the fullest, gave rise to one of the first major theoretical schools concerned with the functioning of mass media, which subsequently developed the hypodermic needle (or magic bullet) theory. Its main proponents, Walter Lippmann (1922) and Harold Lasswell (1938), both argued that the general public was intrinsically unable to form an educated opinion on important matters such as politics, and both called for the employment of directive and corrective propaganda that had to be carefully composed by a governing class. The assumption that Lippmann and Lasswell made was that there is a linear relationship between media causes and effects, between the intention of the sender's message and the interpretation by the receiver. Hence the theory's names: both a hypodermic needle and a magic bullet metaphorically conjure up images of an unstoppable flow of messages, injected in or shot at a passive receiver, who can do nothing but digest these messages exactly as they were intended. In these views, the sower has super seed, and uses it only for what is perceived as the good cause.

Although not everyone believed that the good cause was actually good, the notion that media inject ideas into the minds of unsuspecting people was not uncommon. In the early academic critique on mass media, found most notably in the Marxist work of scholars of the *Frankfurter Schule*, we see a mistrust of the effects of radio and television that is based on a premise similar to that of Lippmann and Lasswell: audiences are passive victims, and are manipulated into becoming part of the culture industries' attempts to sustain capitalist societies by having them consume mass-produced goods.[14] Expressed most vehemently in the works of Max Horkheimer and Theodor Adorno (1947) and that of Herbert Marcuse (1964), the criticism was that we would see a homogenisation of culture, an increasing inability of the public sphere to critically comment on society's values, and ultimately an indoctrination of the people by totalitarian politics. This fear of the power of mass media is deeply rooted in the popular mind; even today, in a time where theorists have long been determined to attribute audiences with far more agency than the Frankfurt scholars did,[15] large-scale events, converging desires, and rampant consumerism, all produced by global industries, still make the Frankfurt School's assessments of the effects of mass media very relevant.[16]

There is another reason why the hypodermic needle theory remains valuable as an example of key theories describing communication as a process in which messages follow a linear and undisturbed path from one source to multiple receivers. Although it was abandoned after the 1950s as an accurate way to model mass media effects, the appeal of the hypodermic needle theory's linearity is still intuitively tempting enough to be invoked for use in strategic argumentation. Partly an expression of folk wisdom and often the result of hasty assumptions about the effect of media on humans, the basic linear principle of the theory continues to rear its head today, for instance in debates induced by moral panic. Typical moral panics concerned with media effects use arguments in the form of 'A causes B', and are often seen in discussions on the 'alleged harmful effects of exposure to popular media and cultural forms – comics and cartoons, popular theatre, cinema, rock music, video nasties, computer games, internet porn' (Cohen 2002: xvii). What matters here is not that these discussions are too superficial, but that their argumentative structures are based on the belief that communication media serve to spread messages, and directly influence the minds of the people that receive those messages.

The underlying myth of communication in the hypodermic needle theory is one that connects the Christian method of dissemination with the intention or desire to have all listening ears tuned to the same fork. In this respect it also resonates with Ogden and Richards' belief that it is possible to minimise the amount of interpretations of communicated messages, and Shannon's desire to ban noise from the communication channel: in the optimal situation there is to be a direct and undistorted mediated transfer of meaning. In the criticism levelled at the supposed effects of mass media, as exemplified by the Frankfurt School and found in moral panics, we read the presence of a media pessimism comparable to the one implicit in the positioning of Socratic dialogue as facilitating ideal communication; mass media are believed to disrupt an authentic state of being together, and to manipulate receivers because of a lack of reciprocity. This criticism, too, however, acknowledges the force with which mass media are supposedly able to supplant the contents of sane minds with foreign thoughts.

The next subsection will turn to idealised ideas of communication that, in some sense, counter these fears of totalitarianism by virtue of seeing everyone as senders and receivers on a global scale, and thus as constituting a grand, interconnected web of knowledge exchange.

Many-to-many communication

Meeting of minds – Pierre Teilhard de Chardin's *The phenomenon of man*: One of the most original thinkers to theorise about 'global connectedness' and its beneficial characteristics was undoubtedly Pierre Teilhard de Chardin, who believed that a continuing convergence of every matter and entity on Earth would eventually lead to planetary peace and unity. His stance was highly controversial at the time of its first reception in the second quarter of the twentieth century, and still is in some respect: as a Jesuit priest and paleontologist, he sought to combine two seemingly contradictory viewpoints on the origin and direction of humankind, namely that of Christian religion (which relies on biblical testimony) and that of evolutionary theory (which relies on paleontological facts). In his quest, Teilhard de Chardin was looking for a way of thinking that could support his conviction that 'the stuff of which all stuff is made [is] reducible in the end to some simple and unique kind of substance' (Teilhard de Chardin 1959: 41).

He found his answer, extensively described in his posthumously published book *Le phénomene humaine* (translated as *The phenomenon of man* in 1959), in the idea that God was not in the heavens above, watching us, but actually inside every organism and object on Earth, including humans. It was this divine essence, he thought, that was steering evolution, and not a secular Darwinian struggle made up by a survival of the fittest and random gene mutations. Moreover, whereas classic Darwinian evolution knows no direction or endpoint, Teilhard de Chardin postulated that the divine essence of life meant that there *was* a direction to evolution, inevitably culminating in what he called the 'Omega Point', a final unity equivalent to 'the new spirit; the new god' (ibid.: 258). The way this zenith of evolution was to be reached lay in a natural succession and transcendent unfolding of Earth's self-consciousness. Elaborating on a categorisation originally proposed by geochemist Vladimir Vernadsky, Teilhard de Chardin saw three stages in the development of the Earth: first, there was the geosphere (the solid crust of the Earth), then the biosphere (the life zone of the Earth), and finally he saw the coming of a noosphere[17] (Earth itself become conscious). The seeds of this noosphere, he thought, were carried by humankind, because 'since the birth of thought man has been the leading shoot of the tree of life', and therefore 'the hopes for the future of the noosphere [...] are concentrated exclusively upon him as such' (ibid.: 276). The continuous growth of interactions between human minds would then ultimately lead to a mega-synthesis in which '*all together* can join and find completion in a spiritual renovation of the earth' (ibid.: 245, emphasis in original).

Teilhard de Chardin's ideas on an emerging noosphere do not so much constitute a theory of communication, as present us with the bold observation that human minds become more sophisticated and connected through time, bringing about a unified consciousness. That this process was both expressed in and was a result of expanding global communication networks was nonetheless very much clear to Teilhard de Chardin; writing in an era when electronic media had started to open up windows to the world, he identified 'the machine' (with which he designated technology in general) as fundamental in bringing about the structure of the noosphere, and pointed especially to

> the extraordinary network of radio and television communications which, perhaps anticipating the direct syntonization of brains through the mysterious power of telepathy, already link us all in a sort of 'etherized' universal consciousness. But, I am also thinking of the insidious growth of those astonishing electronic computers which, pulsating with signals at the rate of hundreds of thousands a second, not only relieve our brains of tedious and exhausting work but, because they enhance the essential (and too little noted) factor of 'speed of thought', are also paving the way for a revolution in the sphere of research. (Teilhard de Chardin 1964: 167)

Thus, for the noosphere to truly come into existence, time, space, and knowledge would have to be linked, and the technological advancements in the fields of communication in the first half of the twentieth century provided just that. For Teilhard de Chardin, the fact that the globe was in a process of becoming encircled with communication cables and radio waves was enough proof that under divine guidance human consciousness had indeed started to expand to global proportions, and in the ultimate end would know only an ideal and blissful state of being.

Another aspect that connects the noosphere with communication theories is that its emergence relied heavily on the direct and unobstructed transfer of thoughts. Emphasising the effects of mass media in a very similar vein as done by the Frankfurt School – but seeing this as a *positive*, instead of a negative, process – Teilhard de Chardin stated that through communication all thought would eventually be harmonised into a 'single thinking envelope', in which the individual mind would be subsumed by the collective mind (ibid.: 251). Also in contrast with the Frankfurt School he contended that this subsummation would not be the result of one-to-many communication; because everything would be connected to everything, all possible thoughts would be broadcasted to all possible listeners and, once internalised undistorted, would form a singular and complete body of knowledge.

Both the noosphere philosophy and the requirements it needs in terms of supportive communication technologies have become very relevant again since the boom of the Internet at the end of the twentieth century. Mainly following up on the work of Marshall McLuhan, who is arguably indebted to Teilhard de Chardin for his identification of electronic communication media as the prime movers of Earth's civilisations towards a global village,[18] contemporary scholars have begun to ask questions regarding the nature of the Internet in terms of creating a 'telepathic society' (Levinson 2004: 57), or becoming an all-encompassing 'sentient environment' (Rheingold 2002: 86). Whether or not the Internet indeed makes up the present-day structural foundation for the noosphere, what remains noticeably visible is that many cyberspace myths recount religious and teleological travels towards a final endpoint.[19]

The communication ideals found in Teilhard de Chardin's vision can thus be summarised in a single mantra: together is good. Divine forces will guide humanity in a converging fashion towards the ultimate transcendent and sublime togetherness, a singularity equal with God. The role of communication networks here is predominantly to facilitate the necessary channels; the mere act of connecting beings and things is considered to provide enough leverage to construct a global consciousness. The content of communication, although not completely irrelevant, gets to play an inferior role. What has rendered Teilhard de Chardin's evolutionary arguments for the creation of a singular super-mind somewhat challenging to follow is that they are heavily influenced by his religious beliefs, but as the next subsection will show, there are secular versions of his arguments to be found as well.

Rising out of chaos – The emergence of a 'Global Brain': The urban anthropologist and self-appointed 'omnologist'[20] Howard Bloom is a prominent candidate for providing an academically informed view on communication and cooperation in biological evolution. As a neo-Darwinist, Bloom's interests lie in explaining virtually everything he encounters in what he calls 'society's myth-making machinery', and while doing so he is able to make grand and sweeping connections between evolutionary phenomena and social behaviour. This is especially visible in his book *Global brain: The evolution of mass mind from the Big Bang to the 21st century* (2000), in which Bloom specifically engages a question that is closely related to the observation made previously: why is it that, while enough evolutionary evidence is at hand to make the claim that a collective mind has already been around for millions of years, only now, with the advent of the Internet, 'an army of equally august specialists' is beginning to warm up to the idea of 'a planet pulsing with a more-than-massive data-sharing mind'? (Bloom 2000: 3). Bloom posits that networked communication has always already been present in evolution, and that it necessarily is the single most important paradigm with which we can understand communication per se.

The keys to Bloom's lengthy and heavily annotated justification for his claim can be found in two propositions. Firstly, according to Bloom, '[f]rom the beginning, we living beings have been modules of something current evolutionary theory fails to see, a collective thinking and invention machine' (ibid.: 2). Here, Bloom follows the point made by Teilhard de Chardin that the seeds of a networked global brain have long been visible in biological evolution. Moreover, he holds that '[n]etworking has been a key to evolution since this universe first flared into existence' (ibid.: 14). Bloom nonetheless distinguishes himself from Teilhard de Chardin – and from contemporary techno-advocates of the global brain philosophy – because he does not think that 'microchips or mystic intervention' are necessary ingredients to ultimately establish a global brain (Bloom 2000: 2). He admits that present-day myths of improved communication in cyberspace will surely heighten our awareness of what global interconnectedness entails, but he argues that in the process of connecting things that has already been going on for millions of years, neither cyberspace nor divine essence is constitutive. In other words, new communication technologies merely open our eyes to what has been happening all along.

Secondly, he argues, group selection is more important to successful evolution than individual selection.[21] It is in every organism's biological make-up to work together and protect the group's interests; connected cooperation wins in the end. The ways in which this networked, connected cooperation sustains itself can, according to Bloom, be grouped into five mechanisms: 1) conformity enforcers (which give the group a shared identity); 2) diversity generators (which do the opposite of conformity enforcers by spawning variety); 3) inner-judges (which judge the actions of their host organisms and give positive or negative feedback); 4) resource shifters (which give to the organisms that contribute to the whole, and take from the ones that do not); and 5) intergroup tournaments (which force each 'group brain' to improve itself in order to survive) (ibid.: 42-44). When viewed as a whole, we can see how these mechanisms form a cluster of elementary evolutionary rules that set up a dialectical motion towards ever-growing interdependency: too much conformity and diversity will kick in; too much apathy and a battle will sort out the weak from the strong. Communication in this process is found at every point where a network of cells, bacteria, ants, cats, apes, humans, societies, or civilisations reshapes itself according to basic rules; the ideal is in its ongoing existence and expansion.

The idea that many small but relatively simple entities can, when combined, form a higher-order intelligence is a principle very much acknowledged in evolutionary biology, where it is called 'emergence'. More specifically, writes Steven Johnson, these sets of rules form 'bottom-up systems', or 'complex adaptive systems that display emergent behavior' (Johnson 2001: 18). Johnson explains that the process of self-organisation among lower-scale agents is a recurrent phenomenon, visible in the beginnings of life as well as in brain-simulating software. Here, Johnson shares the argument put forward by Bloom that networking and therefore intercommunicating capabilities have always been part of biological make-up, and as such are not specifically a consequence of modern communication and computer technology. However, Johnson does want to investigate the particular role of those technologies, and engages the myths that cyberspace itself shows signs of developing a global consciousness in a less dismissive way than Bloom does. Referring to Robert Wright's *NonZero: The logic of human destiny* (2000) – in which Wright presents a view on the transhistorical nature of self-organising and emergent principles similar to that of Bloom and Johnson himself – Johnson writes that today it is not as strange as Bloom makes it out to be to argue with Wright that '[m]acrointelligence emerged out of the bottom-up organization of city life [...] and it will do the same on the Web' (Johnson 2001: 116).

Despite its ability to connect things like never before, says Johnson, there is also a very important reason why the Web itself will *not* automatically morph into a conscious global brain: 'intelligent systems depend on structure and organization as much as they do on pure connectedness' (ibid.: 118). He proceeds by demonstrating that the Web knows no structure similar to that of a city or a brain; it merely connects, and in contrast to Teilhard de Chardin's view, this is not sufficient. What is needed to obtain an orderly structure is, according to Johnson, a feedback mechanism, and 'given the Web's feedback-intolerant, one way linking' (ibid.: 121), a global brain will not spring out of merely interlinked channels of communication. Only if artificial adaptive systems adhere to the basic rules of emergent behaviour will there be a chance for a higher-order intelligence to arise.

Looking at the communication ideals found in their work, we can see that Bloom and Johnson emphasise that the communicative processes that form, maintain, and grow networks are 'quasi-natural' processes, ones that have been with us from the beginning of time. The idealness of these processes is already implied in their functioning; there is nothing that can fundamentally disrupt them, because they constitute an adaptive system that will automatically sort out problems to find its new and improved synthesis. This vision of how an ideal state of being is already part of nature echoes Teilhard de Chardin's views: just let the simple, basic rules do their work, and communication will inevitably improve itself so to accommodate ever-growing connectedness. The problem with both Bloom's and Johnson's accounts, however, is that they tend to discard human agency in giving a particular shape to that connectedness. Humankind's contribution in bringing about a global consciousness is nil if we follow Bloom, and marginal at the most if we follow Johnson. The next subsection will look at how a more dedicated perspective on human agency is presented in Pierre Lévy's view on global interconnectedness.

Reinterpreting the Internet – Pierre Lévy's *Collective intelligence*: The wrapping of the Earth in multiple layers of interconnected electronic communication networks, which started in the eighteenth and continues into the twenty-first century, has given validity to the question whether we have indeed become part of the establishment of a higher order of intelligence, as Teilhard de Chardin believed. Although an increase in connectedness, seen from the views of both Bloom and Johnson, is nothing new in evolutionary biology, it was only when people started to surf the World Wide Web and contribute to its informational growth that the question of the emergence of a global brain truly began to register among media theorists.[22] By dismissing cyberspace as simply a footnote in the history of connecting things (as Bloom does) or by focusing on how the Web *itself* can become conscious (as Johnson does), there is a risk of decreased attention for how particular and increased human interactions, mediated by communication technologies, might harmonise into something like a universal understanding. It is at this point that philosopher Pierre Lévy's (1997) imaginative notion of 'collective intelligence', located in what he calls the 'knowledge space', presents a solution.

Lévy describes the evolution of earth-bound living as a process that has known a succession of four types of space, in which the organisation of knowledge and information is determined by the tools and symbols predominantly available in those spaces. The first three spaces Lévy defines as the nomadic, the territorial, and the commodity spaces. Here, respectively, Lévy sees humankind tell myths and perform rites; develop writing, geometry, and cartography; and build a global economy of 'material and statistical goods'. The knowledge space is the fourth space, the one we are now starting to inhabit. This space constitutes an 'informational cloud', a 'space of living-in-knowledge and collective thought' (Lévy 1997: 140-141). It is in this cloud that Lévy locates cyberspace, and in which he identifies the possibility for the emergence of an 'economy of knowledge' (ibid.). An important premise for the existence, growth, and preservation of the informational cloud is that all people interact with it by adding, changing, and retrieving data in whatever way possible. As such, and very much resonating with Teilhard de Chardin's notions of the noosphere, this activity of global communication will 'unfold and grow to cover an increasingly vast and diverse world' (ibid.: 111-112), ultimately creating a universally accessible information realm.

Lévy's hierarchical description of the evolution of knowledge storage and transmission reverberates with Albert Borgmann's (1999) historical analysis of the nature of information, in which he makes a distinction between propositional knowledge (information about reality), procedural knowledge (information for producing and realising reality), and virtual knowledge (information as reality). Both Lévy and Borgmann present us with historical shifts in the dominant forms of information, in which that information is being ever-more dematerialised through the use of new communication technologies. They differ, however, in their opinion on whether this is good or bad; while Borgmann warns of a potentially dangerous split between information about/for reality and information as reality,[23] Lévy does not seem to believe there is any risk of leaving material reality behind. He frames the knowledge space firmly within the other three spaces: '[It is n]ot exactly an earthly paradise, since the other spaces, with their limitations, will continue to exist. The intention of collective intellect is not to destroy the earth, or the territory, or the market economy' (Lévy 1997: 141). For Lévy, it is clear that a collective intelligence, virtually established in the informational cloud that is cyberspace, presents us with a progressive, and not a regressive, reality.

Because the first three spaces are not obliterated by the arrival of the knowledge space, in a circular movement – 'a return of the earth to itself' (ibid.), as Lévy calls it – the knowledge space connects back to the other spaces. As such, a spiral pattern emerges: old habits and identities remain present in the structure of life, but as a whole this structure is raised to a harmonious state in which more and more people will come together, all through the sharing of knowledge. Traces of the search for a perfect language can be seen again as well: central to the success of the constitution of a knowledge space, according to Lévy, is the creation of a set of signs and symbols that is contributed to by anyone and accessible to everyone. Indeed, Lévy sees that these languages 'become ever more interconnected' and that as such they present us with what he calls the 'surrealist mirror-image of the collective intelligence' (ibid.: 197).

Whether it constitutes an earthly paradise or not, Lévy cannot help but describe the knowledge space in terms of a utopia 'waiting to be born', 'a cosmopolitan and borderless space', 'an electronic storm', and 'a sphere of artifice shot through with streaks of light and mutating signs' (ibid.: 138-141).[24] There is undeniably a transcendental element visible in the way Lévy writes about the knowledge space, in which information is to be uncoupled from its static bases. This dematerialising movement fits perfectly with the religious undertones found in other myths of networked communication: liberating things by releasing them from their carriers promises more opportunities to interconnect those liberated entities, as they have become free-floating nodes in a dynamic network. Thus, in the end, in its most radical form, the idea behind a collective intelligence – one that is brought forward by people's creative and straightforward use of cyberspace's connections and freedom of expression – is that each node can be directly or indirectly connected to all others, providing instant and complete transferral of whatever form of information. Lévy stresses that this information always constitutes meaning, for all users of cyberspace will gather knowledge of all existing signs and symbols, simply by their online presence. The idealised idea of communication found in Lévy's collective intelligence is thus that mere participation in a network will eventually result in a global and shared production and transferral of knowledge.

The conceptions of many-to-many communication, as visible in the works of Teilhard de Chardin, Bloom, Johnson, and Lévy, display a range of various perspectives on what the driving forces behind the processes of interconnection are (be they divine powers, quasi-natural evolutionary forces, or the combined actions of people collecting knowledge and handling information). What all of these perspectives have in common is the premise that the formation of networks, in which multiple and dynamic connections ensure the spread of information and knowledge to multiple nodes, is generally a good thing. Their underlying idealised ideas of communication aim to bring about and maintain those networks, so as to establish in the end an ultimate network of networks in which all nodes are interconnected.

Taking the analyses of all three forms of communication outlined in this section together, we can distinguish different basic notions of what communication is, and different strategies that describe to what extent it can, should, or will be improved in order to eventually arrive at a communication utopia. There is the idea that true communication is about unifying channels, or about synchronising thoughts. Another approach to improve communication is to have everyone speak a shared language, and yet another is to just have everything connected to everything in a quasi-naturally evolved all-encompassing network. While there are many nuances in these ideas of communication improvement, as well as in the underlying dissatisfactions from which the ideas originate, they share distinct similarities with the reasoning behind the utopian projects analysed in chapter 1. What they have in common is a desire to mend conflicts – *any* conflict – by trying to establish an ideal situation in which there can hardly be a conflict at all. As was the case for utopian societies, the implication is that if every act of communication is to be a seamless and conflict-free process, this then calls for a universalising approach to the organisation and functioning of communication. We saw that this type of reasoning is not only impossible without running into paradoxical situations, it actually ignores the contents of communication processes by mostly stressing their phatic function only, and denies the humanness of not being continuously in rapport. Remove the cause of conflict, and what is left is a serene but zombie-like bliss (Crombag & Van Dun 1997: 49). So why even ponder this tragic desire to remove communicative noise?

The importance of noise – and its repression

> *[N]oise is the interference that is simultaneously disruptive and creative. There can be no information without noise and vice versa. Noise can no more be silenced in the world than parasites can be exterminated; life depends on parasites as much as information depends on noise. (Mark C. Taylor 2001: 121)*

The focus so far in this chapter has been on how regularly returning discursive elements in the search for optimised communication invariably contain the idea that there exists a communicative framework that encompasses all ideals. We saw that, although the ideals which were discussed allowed for various degrees of interpretation of their ontological status and differed in the way they could be fulfilled, their common denominator was that they were *there*, linked as interrelated family members born out of the communication sublime. Each exudes a sense of fulfilment, of complete togetherness, of utter understanding and of conflict-free being-as-one. In this sense, myths of ideal communication share many utopian narrative strategies; moreover, there is a lot to suggest that utopian stories that depict a perfect society by default presuppose that there can be no misunderstanding, not of the ground rules which constitute and uphold that society nor between its inhabitants in their day-to-day encounters. Perfectionism, after all, does not tolerate deviant behaviour or ambiguous interpretation.

It is therefore both fascinating and troubling at the same time to see that universalising communicative blueprints have, for so long, held a firm grip on the human imagination – and judging from the recurrent mantra in ads for mobile communication technologies, still continue to do so. While there is clearly a positive psychological function to be found in the promise of never again having to experience frustration in understanding one another, the ultimate fusing of minds at the same time requires the playing field of communication to be evened out to such an extent that the communicating subject loses its 'self', and becomes subsumed by an anonymous cloud of pure and continuous understanding.[25] Referenced in dystopian and anti-utopian narratives and echoed in the critique of idealised models of communication, the price paid for the human wish to eradicate disruptive noise from universalising processes is a loss of personhood, and hence a negation of that which makes us human in the first place.[26] A paradox! Thus, in order to provide a better understanding of what communication is when it is not being idealised (and why people think or say it should be), we need to examine the role and function of communicative noise.

Celebrating noise

Although a large portion of attention so far has been given to the overwhelming power and pertinence of myths that tell of overcoming obstacles and reaching a final and fulfilled state, this does not mean that these myths dominate the field of communication theories. The notion that communication loses its meaning when there are no hindrances in the communicative act has not been lost upon communication scholars – let alone the platitude that what is ideal for one person can be a nightmare for the other. To communicate is to struggle, to acknowledge the differences between the communicating parties, even to misunderstand. Failing to recognise the fundamental role of noise, the 'messiness of life', is to see communication from a particularly unproblematic and instrumentalist perspective, as Kaustuv Roy notes: 'It is when we take this messiness of lived life seriously, the "noise" that techno-rationality seeks to brush aside, communication appears much more problematic than previously imagined' (Roy 2004: 299). When the constitutive function of noise is taken into account, communication thus becomes something quite different from the perfect transparent transmission of meanings-in-themselves, and stops being the mythically envisioned tool to bring about a sublime togetherness.

Communication theorists generally distinguish four types of noise: mechanical (or physical), psychological, cultural, and semantic noise (DeVito 1976). The first type, mechanical noise, is present at the physical level and manifests itself as breakups of a signal: hisses on a telephone line or any other interruption or distortion of a flow of data. This is the type of noise that Shannon was primarily concerned with, and he succeeded in filtering it out of the communicative process. The second type, psychological noise, occurs when the psychological state of a person is such that it influences the interpretation of a received message in an unexpected, skewed, or biased way, for instance when a recent widow is inadvertently told that the hearse carrying her late husband will arrive 'dead on time'. The third type, cultural noise, is in some respect similar to psychological noise, in that it, too, involves biases and as such is visible in situations where culturally determined messages are understood differently by people who are not familiar with that culture or have a certain bias against it. Both psychological and cultural noise appear as nemeses in the therapeutic and global brain discourses in idealised communication, like those of Ogden and Richards or Teilhard de Chardin: if both types of noise are transcended, the argumentation goes, there is nothing that stands in the way of truly understanding each other, irrespective of one's culture, and letting us glide into a perfectly peaceful world. The reality is of course that both psychological and cultural noise are hardwired into our being and upbringing. The fourth and final type, semantic noise, is the type that is connected to ambiguities present in language, stemming from the ways in which multiple meanings can be attached to the same signs and vice versa. Again,

Ogden and Richards give a good example of wanting to tackle this type of noise, and also Lévy is known to have tried to extensively map all semantic codes in cyberspace;[27] tragically so, however, because semantic noise not only produces ambiguities but also lies at the root of irony and paradox, both of which, since Gödel, have been acknowledged as inherent to any symbolic system, and cannot be ironed out.[28]

So of the four types of noise mentioned above, only the first, mechanical noise, can said to be actually suitable for elimination by technical means.[29] What is striking, however, is that in myths of ideal communication, the other three types are often approached as if they were merely disguised versions of the first type, which opens up the idea that 'the imperfections of human interchange can be redressed by improved technology or techniques' (Peters 1999: 29). Chapter 3 will give a more extended historical overview of such discourses of communicative improvement through technological means, but for now it suffices to note that the reason the communication sublime has continued to perpetuate myths of ideal communication over time is to be found in the eternal presence of noise *in general*. Noise engenders the longing for a communicative utopia, it is the necessary opposite, anti-perfect form of communication; more specifically, and recalling the language of Bloch, communication can only be defined in relation to non-communication, as it expresses a longing that can only exist when there is something not-yet-existing. As a complex system of signs, language, and meaning, communication is inherently dependent on noise in order to exist and survive at all; like an oyster produces pearls when its shell is infiltrated by irritants, people produce communication when their life is confronted by disorder.

In understanding this notion of communication as a process fundamentally revolving around the presence of noise, a useful perspective comes from the work of Sören Kierkegaard, the Danish philosopher who is well known for his attacks on ideas of direct, transparent, and unmediated communication. Living at a time when Hegel's writings on the attainment of absolute knowledge dominated philosophical debates in Europe, Kierkegaard sought to debunk the claim that pure thoughts could be dialectically made available.[30] The key to this firm stance against communication as the neutral vehicle of objectively shared messages is found in *Frygt og bæven [Fear and trembling]*, in which he presents his critical reading of the story of Abraham's offering of his son Isaac. Here, Kierkegaard – or rather his pseudonym Johannes de Silentio – highlights Abraham's act as utterly incomprehensible, and asks how it could ever be understood when it is precisely the point of the story that some things are simply incomprehensible. If every witness would have known of God's command to Abraham, the act would have immediately lost all its demonstrative power. Abraham cannot communicate his thoughts, because that would undermine the essence of his faithful deed, Kierkegaard/Silentio writes: 'The ethical expression for what Abraham did is that he intended to murder Isaac; the religious expression is that he intended to sacrifice Isaac. But in this contradiction lies precisely the anxiety that indeed can make a person sleepless, and yet Abraham is not who he is without this anxiety' (Kierkegaard 2006: 24).

Kierkegaard argues that Abraham is who he is, exactly *because* his thoughts and acts are not understood; they are *supposed* to be fundamentally out of reach, for that is what marks him as a man of flesh and blood, a particular instead of a generic person. Kierkegaard provides us with a very useful way to think about communication as a private undertaking that can engender mixed interpretations, as opposed to it being an open and public process that should be readily understandable by anyone. This distinction between the private and the public runs through many of Kierkegaard's writings, and it not only underlines his critique of Hegelian philosophy; according to Peters it also emphasises '[h]is insistence on singularity rather than generality as the ruling principle of our relations with each other' (Peters 1999: 134). What we should take from this, is that the tension between what is good for the individual and what is beneficial for the group is at the root of many of the problems that arise when forging idealised ideas of communication. There is a notorious danger in thinking that communication is ideal when it has become generalised and universal, a danger that has been identified earlier in the analysis of utopian narratives and one that Kierkegaard as well warned us about: we can *think* the universal ideal, but we exist in a particular, lived moment, and there is no way that those opposite polarities can be reconciled. Easy communication with others does not exist, and it is a mistake to think it can or should be made easy by trying to eliminate what is perceived as unwanted noise.

Communication from this perspective is an activity that, while trying to bridge the gaps between minds, continuously has to deal with facing the otherness of 'the other', without ever coinciding with it. This understanding of communication spans the broad existentialist movement following Kierkegaard,[31] and has informed a range of philosophical enquiries into the nature of communication in the twentieth century. We see Martin Heidegger (1927) stress that expressing the differences in being-with-others is precisely what makes us human, and Karl Jaspers (1935) argue that it is only by continuously reinterpreting the world that we experience our authentic existence. Also in Gilles Deleuze and Félix Guattari's poststructuralist explorations of language we find similar non-instrumentalist notions that celebrate communicative noise and shun the idea that communication should or could lead to a perfect one-to-one understanding or even a grand unified togetherness. In analysing language as indirect discourse in *Mille plateaux* [*A thousand plateaus*], they define noise as a necessary element of language, breaking down the dialectical opposition between meaningful information and meaningless noise (Deleuze & Guattari 2004: 87-88). Thus, in multiple approaches towards understanding communication, the notion can be observed that in the act of facing others and trying to convey meaning, there can be no single measure of the trueness or purity of a transmission, as all utterances are equally part of everything that can be said.

To sum up, myths of ideal communication fly in the face of lived experience, that is, they tend to ignore the fact that to struggle *is* to communicate; transferring or sharing things takes effort. While the notions of improved communication described earlier focus on how individuals or groups of individuals can be unified in a single understanding, the more experiential view on communication reminds us that we are individuals precisely *because* there is misunderstanding. Imposing a totalitarian scheme upon communication will result in the same problems as those found in attempts to realise utopian projections. The remarkable thing about this is, of course, that ideas of idealised communication retain their seductive powers of promise – even when we are aware of the grave consequences of fulfilling those promises. We have seen this paradox again and again in history, and we continue to see it, especially in contemporary discourse on communication media in general, and on wireless technology in particular. This then calls for an investigation of the ways in which the promise is upheld, by whom and for what purpose.

The art of repressing noise

It should now be clear that idealised ideas of communication share the same double-edged relationship to paradoxes as utopian narratives do. They arise from a central paradox in our eternal hopeful quest for purpose, while simultaneously, when pursued, they create and sustain further paradoxes. The communication paradox is that, in the desire to ban noise from communication and create an all-encompassing sea of understanding, there is at the same time a push for losing that very element that makes communication what it is. Realised utopias and pure understanding end up being experienced as either nightmares or seas of ignorant and zombie-like bliss; in both cases we cease to be human in the sense of what we currently experience humanness to be, *including* the act of desiring pure communication. So, as was the question with utopian narrative, why desire such a state at all? Why advocate the idea that communication inherently *suffers* from noise, and that it should be improved by eliminating noise altogether?

To answer these questions, we only need look at the ways in which myths of the communication sublime are translated into concrete motives for communicative strategies in everyday life, and how these are used by social actors in various (cultural, commercial, political) settings. The motives that arise from bodies of fiction and myth, after all, are not neutral. People have all kinds of goals in their lives, and the ways in which communication can be used to achieve those goals will more often than not collide with how others understand what communication is for. Following literary critic Fredric Jameson's description of shared ideas of progress as a 'vaster narrative movement in which the groups of a given collectivity at a certain historical conjuncture anxiously interrogate their fate, and explore it with hope or dread' (2005: 282), motives for interpreting myths of the sublime in specific ways are informed by a multitude of social and historical factors. In gauging how master narratives of communication bliss are appropriated to steer both individual and public expectations towards hope or dread, therefore, it is vital to recognise that attempts to isolate key elements for improvement also strategically set agendas for reinforcing underlying ideologies.

In the case of communication, ideologies of hope and dread revolve around the issue of noise, whether it should either be abolished or upheld to improve communication. In both cases, the right action clearly depends on what images of the future are entertained, and, more importantly, by whom these images are produced. If the future of communication is presented as a world consisting of uninhibited understanding, noise is what should be taken out of the equation; if it depicts human uniqueness as the highest good, noise is what makes that uniqueness possible. What often tips the balance in favour of cancelling out noise in communication ideals is that the two visions of the future have, over time, not received equal attention. As the previous chapter showed, especially since the Enlightenment and during the development of communication technologies in the nineteenth and twentieth centuries, there has been an ever-increasing emphasis on working towards futures that would be efficient, would know no wars or violence, would provide ample goods for everyone, would promise equality and, above all, would realise a Universal Brotherhood of Man. What becomes apparent when dissecting these futures is that clear communication is invariably seen as a key in manufacturing their success; only through true and complete understanding can universality be attained. This idea is so powerful that it continuously sustains itself, even when lived experience favours images of a future that incorporates noise. As Peters writes of the deceiving lure of the idea of clear communication: 'It invites us into a world of unions without politics, understandings without language, and souls without bodies, only to make politics, language, and bodies reappear as obstacles rather than blessings' (Peters 1999: 30-31). Thus, in a now-familiar paradoxical fashion, plans to reduce communicative noise in visions of ideal futures always inadvertently generate new obstacles, for which improved communication again seems to be the answer.

This power of the notion of communication as something that, when it is noiseless, is capable of finally bringing about an earthly paradise, has not been lost on social actors involved in the act of future-making. It has proven to be a very effectual strategy to stress that, precisely through obstacle-free communication, all kinds of doors will open that were previously locked because of inadequate understanding. In this view, communication works like a metaphysical crowbar, bestowed with no particular function other than to catalyse improvement, and as such it carries what historian Leo Marx would call a 'hospitality to mystification' (Marx 1994: 249). Writing about the word 'technology', Marx shows that its relative abstractness has 'a kind of refining, idealizing, or purifying effect' so that it 'invites endless reification'(ibid.: 248), and it is clear that the word 'communication' also fits this characterisation very well. Communication works as an 'ideograph', a discursive building block that rhetorician Michael McGee defined as

> a high-order abstraction representing collective commitment to a particular but equivocal and ill-defined normative goal [that] warrants the use of power, excuses behavior and belief which might otherwise be perceived as eccentric or antisocial, and guides behavior and belief into channels easily recognized by a community as acceptable and laudable. (McGee 1980: 15)

People who produce predictions about the future make very good use of the fact that the lure of a better tomorrow, be it through the so-called improvement of communication or technology (or both), thrives on the powerful grip that these particular ideographs have on our imagination. Future images that do not incorporate other ideographs such as 'peace', 'democracy', or 'equality' are not attractive, and in contemporary culture they are even less so when they omit communication technology as the primary means with which they are to be realised. For this reason, we can see that most of the concrete motives to improve things are used by social actors who are active in the field of producing and marketing communication technology.

Invariably, these social actors employ ideographs in two specific strategies, aimed at portraying the future in such a way that it stimulates an urge to improve things. Philosopher Rein de Wilde (2000) identified these strategies as invoking either a 'beckoning future' or an 'onrushing future'. The beckoning future presents us with the promise that anything is possible, owing to the prospect of radical new beginnings. It is often cosmopolitan, De Wilde observes, in that it advocates breaking down boundaries between countries and cultures and promotes the building of a tolerant world society (Wilde 2000: 88). This is an image we recognise in the revolutionary aspects of utopian narratives: a desired break from the present is projected upon the existence of a radically new and different place or time, which is glorious and knows only happiness. The gist of beckoning future stories is that it might not be straightforward to reach that place or time, but it can be done; De Wilde points out that most prophets of the beckoning future do acknowledge that the road to the promised land can be full of risks, but by reframing these dangers as 'challenges' that can be overcome, the prophets are able to brush aside any obstacles (ibid.: 82). The catch is, of course, that in the present time we need to establish the conditions under which the beckoning future can become reality, and this is the point where ideologies and ideographs come into play: improve technology, improve communication and use it efficiently, and the future will then unfold itself as promised.

The second form of future-making can be traced in stories of what De Wilde calls the onrushing future. As its adjective suggests, compared to the beckoning future, this type is at the opposite end of the spectrum of the not-yet-become, telling us that it will inevitably engulf us. There is no escape, no chosen road towards the promised land, the future itself will have arrived before we know it. De Wilde notes that onrushing futures always present themselves as already fully shaped, highlighting problematical differences with the present and therefore urging us to prepare ourselves *now* or run the risk of being 'wiped out' (ibid.: 86-87). A typical example of how this logic works is found in the prophesied coming of the 'knowledge economy', which on a global scale imposes the need to invest in information and communication technologies. Instead of working on the conditions that will make possible the future as we would like it to be, we are panicked into thinking that it might be too late for making certain choices, if any at all. Thus, De Wilde adds, stories of an onrushing future are often framed as a contest in which there can be only winners and losers; the mantra tells us to keep on doing what needs to be done, 'or else...' (ibid.). And again, it is evident that the right courses of action in these cases are in no sense neutral. They are constructed by those who have invested in certain future scenarios, by industry lobbies or special interest groups, by various political forces, and by science and technology policy programmes. The resulting agendas, conditions, and priorities typically leave out alternative or unwanted views, thereby emphasising that the future is of a singular and unobstructed nature.[32] Reiterating the example of the knowledge economy, an ideograph such as communication subsequently plays a large role in these singular and unobstructed futures, as its attractiveness as an umbrella term for anything that has to do with 'removing obstacles' automatically places a great sense of urgency on the need to extend and improve it.

To recapitulate, the art of repressing noise thus works on three levels. First, on the level of speech acts, communicative noise (be it technical, psychological, cultural, or semantic) is experienced as disturbing the transmission of the very ideas, mindsets, regulations, and actions that make it possible to bring about a better future in the first place. Ergo, the instrumental thinking goes, the less communicative noise there is, the closer we get to an improvement of communication. Second, projections of the future work because they either stress the importance of removing obstacles on the road to an earthly paradise, or warn us that if we do not prepare ourselves by removing obstacles from the scene where the future will land, we will be cut off from partaking in that future. In both cases, whatever is represented as an obstacle is seen as the noise that needs to be eliminated in the channel that links the present with the future. Third, and in its most general sense, repressing noise is part of the intrinsic logic of future-making. As De Wilde demonstrates, both beckoning and onrushing images of the future function in a similar way, because although they each might have a different point to make, they both typically pay no attention to the extremely varied (noisy) consequences of their respective programmes, and instead focus on clear-cut and singular (noiseless) outcomes (ibid.: 95).

Finalising the analysis of the function of noise in communication, the conclusion must be that it has an ambiguous role. On the one hand, noise is placed at the root of communicational problems, which subsequently fuels the urge to translate idealised ideas of communication into strategies that suppress noise and, as such, promise to bring forward a better future or prepare us for its arrival. On the other hand, noise can be regarded as the fundamental reason why communication exists at all; just as happiness in an achieved utopia will have lost its meaning, communication without noise would cease to be the characteristic that defines us as part of a globally interconnected web of unique human beings. Noise is the quirk, the necessary paradox in communication, and any amount of trying to eliminate this noise will just ensure that the attempt remains an impossible dream.

And yet, we continue to try. It is like the wish to communicate as angels, who 'either understand the thoughts of others, or [...] can read them in the divine mind' (Eco 1995: 37), and it is a wish that continues to haunt us. As Peters notes, angels

> provide us a lasting vision of the ideal speech situation, one without distortion or interference. Angels – a term that comes from the Greek angelos, messenger – are unhindered by distance, are exempt from the supposed limitations of embodiment, and effortlessly couple the psychical and the physical, the signified and the signifier, the divine and the human. They are pure bodies of meaning. (Peters 1999: 74-75)

The next part of this book will therefore turn towards the role of communication technologies in the many expressions of such dreams of angelic communication and investigate the numerous technologically driven endeavours in history that have tried to solve the perceived problems of communication, from the telegraph to the present-day mobile communication device. Again, we will encounter the necessary fictions that tell of how improvement will enable us to reach the utopian communication sublime.

Part II

Where Angels Speak

3. The rise... and rise of media technology

Someone had to observe that tapping with an electric finger would stir vibrations over a far-flung range beyond the orbit of the moon. Genius had to discover in the Creator's plan a mysterious medium hidden until the twentieth century approached, when mankind's accelerated pace needed communication on a world-wide scale; more rapid than the mail carried by an ocean liner, faster than an airplane dispatch, quicker than the telegraph of Morse or the telephone of Bell. (Orrin E. Dunlap 1937: 38)

So far in this book, the focus has been on how the search for hope and purpose and its accompanying narratives create progressive and utopian agendas, which subsequently inform the creation and preservation of idealised yet paradoxical ideas of communication. It is now time to see how this all relates to technology, and how we can understand it as a myth-oriented domain in its own right. That is, we will tap into the fantastic discourses that have continuously surrounded communication technologies, regenerating and maintaining the idea that new media bring us closer to truly coming together and understanding each other. This chapter is about our fascination with the transcending experiences that new communication technologies continuously seem to promise, or in the words of media historian Jeffrey Sconce, about why it is that 'after 150 years of electronic communication, we still so often ascribe mystical powers to what are ultimately very material technologies' (Sconce 2000: 6).

First, since myths of ideal communication have been proposed to be regulative principles, it is necessary to question which theoretical and conceptual tools are best suited to study how myths can be considered as 'guiding forces' in technology development. After contrasting the viewpoints of technological determinism with those of social constructivism, I will suggest that both perspectives only limitedly support opportunities to investigate the agency of myths, and that this is more or less also the case with alternative views like those from actor-network theory and 'symptomatic' approaches. Next, I present the approach to describing media history with which I consider it to be possible to acknowledge the contingency of the development of communication technologies, while at the same time recognising the regulative function of idealised ideas of communication. This approach is based on an evolutionary model of technology development, in which directive principles are assumed to be present without having to accept teleological explanations, and with which a multitude of environmental factors can be described that play a role in constructing technologies, including forceful fantasies and moving myths. This approach also provides a valuable way to stress the important difference between the desire to *see* progress in media development, and *describing* this desire as a recurrent element in media history.

Finally, I give an overview of myths of ideal communication, contextualised by the heterogeneous environments in which they manifested themselves in media evolution, and show that their manifestations are intrinsically linked with the often strategic need to endow new technologies with the most fantastic qualities. Grouping the multiform communication technologies into five 'constellations' – the telegraphic, telephonic, radiographic, televisual, and Internet constellations – I aim to demonstrate that, while communication media are regularly placed in a rigid and sequential progressive line, in reality they know no such clear-cut boundaries, and are all products of often simultaneous and evolutionary processes in which ideas of communication utopia are abound.

Valuing myths

Myths are the most powerful moving forces behind technology.
(Lee Worth Bailey 2005: 32)

A common element in the theorising of the idea of progress is that its birth is intricately linked with technological advancement from the eighteenth century onwards. Although this view has considerable merit, it needs to be refined; as was demonstrated in chapter 1, perceptions of linear time, and thus of ways to place events along a path towards a certain goal, can be distinguished as early as in ancient Greece. Even so, we also know that, at that time, linearity was enclosed by a dominant circular mindset, and that, while we might be tempted to retrospectively locate current understandings of ideas of progress in classical times, it is a mistake to think that there could be a single, univocal, and a-historical perception of what progress entails. Myths of progress, seen as an upward movement feeding on ever-present notions of 'newness', have gone through many transformations, especially regarding the nature of foreseen goals.

What we can say is that the idea of progress in its now-familiar version enjoyed its most dominant phase during the Enlightenment. In that period, a shift took place towards seeing reason as the primary means to bring about progress, mostly under the influence of what can be called the 'new science', a way of thinking progressively whereby 'each generation start[s] with the best of the last generation, and [is] bound to add to it' (Pollard 1968: 6-7). But not only those working in science laboratories came to know of new developments, as discoveries were moved more and more into practical everyday use. As Leo Marx writes, '[n]othing provided more tangible, vivid, compelling icons for representing the forward course of history than recent mechanical improvements like the steam engine' (Marx 1994: 240). By their growing visible presence, technological artefacts gave rise to the idea that nature could be conquered, and that man could gain control over his destiny. Especially the advent of industrialism had a huge impact on the social and cultural lives of many people, fuelling progressive myths that standards of living would be raised universally.[1] Increasingly, new technologies manifested themselves in the collective consciousness as the single-most notable sign that humanity was moving in the right direction, thus strengthening the myth that they, and they alone, would bring about the long-awaited earthly paradisiacal state (Noble 1997: 4).

Hence, ideas of progress have often been intimately tied with a deterministic take on technology's influence.[2] Stock examples of how this intertwined relation is perceived tell of how 'technology X changed the world for the better', or that 'technology Y finally solved problem Z'. In each case, a direct link is established between a technological innovation and a positive effect, suggesting that there were no other factors involved. Such myths are still with us today, and remarkably so, considering that the notions that technology is the prime mover of change and that improvements in technology automatically entail progress have, mainly in the twentieth century, received a fair share of criticism. World wars, atomic bombs, technological disasters, and global warming may have tempered the widespread belief in progress through technological means, but its momentum appears not to have faded significantly, especially if we look at how new communication technologies are still presented and perceived as able to provide solutions to everyday and complex problems. The question here, then, is how we should value the functioning of myths in our perception of the relation between technological development and social change. We have seen earlier in this book that myths are not simply ornamental stories, but what exactly is their status when technology is concerned? Should we conclude that, because they are ever-present, myths greatly inform – perhaps even determine – the development of technology, or is their presence more of arbitrary influence? How can we conceptualise their role? To answer these questions, I will briefly discuss the set of approaches with which technology is generally studied, and then assess to what extent these approaches provide the necessary tools and opportunities to ascertain how myths of idealised communication play a part in shaping communication technologies.

When analysing the relation between technological developments on the one hand and social change on the other, most scholars generally distinguish two distinct views, that of *technological determinism* and *social constructivism*. Sociologist Claude Fischer provides a useful illustration of the broad contours of these two views, and, as will become clear, the means with which to locate the position of myths in them as well. First, Fischer discusses the notion of technological determinism – or what he calls the *impact analysis* approach – and from the outset it is clear that determinism is a controversial subject. Fischer shows that the idea behind impact analysis is often explained with the metaphor of a billiard ball effectuating a force upon other balls on a billiard table, scattering them according to natural laws and generating a cascade of 'impacts' until the strength of the initial force has subsided (Fischer 1992: 8). Technology in this model is entirely external to society, and, when it at some point enters that society, it does so autonomously, and with predetermined consequences. As an example, Fischer presents historian Lynn White's argument that 'the invention of the stirrup led, by a series of intermediate steps, to feudalism' (ibid.). This view on technology can be characterised as making use of a macro-scale perspective on history, relying on assumptions of linear monocausality, and ignoring technological side effects and coincidences. The 'softer' version of technological determinism, which Fischer calls the *impact-imprint* approach, does recognise the contingency of technological developments, but also finds causal effects in how new technologies culturally and psychologically 'imprint' their essential qualities on their users (ibid.: 10). Here, Fischer refers to scholars such as Stephen Kern and Joshua Meyrowitz, who argue that time- or space-transcending technologies necessarily engender widely held senses of 'timeless-ness' or 'placeless-ness' (ibid.: 10-11).

The second approach to studying the relationship between technology and society is one that Fischer identifies as *social constructivism*. Following the founding work of Wiebe Bijker and Trevor Pinch, this approach sees technology as an undetermined heterogeneous field, in which struggles and negotiations between diverse social groups shape particular structures and definitions of particular technological devices. Such a perspective would be best suited to give an adequate explanation of, as Fischer writes, why 'the same devices may have different histories and uses in different nations' (ibid.: 16). It would also take into account the role of end-users in co-shaping technologies. Fischer places a great deal of emphasis on the need to study people's agency and the limits of that agency in dealing with technology, and rightly so; it would be nonsensical to claim that humans are subservient creatures, all volitionally manipulated into using technologies the same way and exactly as was intended. To attribute primary agency to technology is to ignore socio-political struggles, economic factors, and empirically grounded evidence that technology rarely has a single use or effect, all of which *are* acknowledged by a social constructivist approach (ibid.: 17-18). Studies that zoom in on the micro-scale of technology development, Fischer contends, are far better at avoiding familiar deterministic traps, such as focusing on successes only, or wanting to distinguish between 'true' and 'false' origins of certain artefacts.

Of the two approaches, technological determinism is clearly not favoured by Fischer, and he voices a widely held view.[3] He dismisses it for being too reductionist, and for ignoring evidence that many technological artefacts are produced by particular social groups for particular purposes; as he subsequently notes, often technological developments could have ‘unfolded otherwise were it not for social, political, or cultural circumstances’ (ibid.: 9). These are important objections, as they also largely hold true for the project of determining which approach provides the best opportunities to study the role of myths in orienting technological development. If technological determinism portrays technologies purely as rationally derived tools that, in their value-free neutrality, develop autonomously and are impartial to social, ethical, or psychological factors, it leaves very little, if any, room for studies of how non-rational hopes and dreams might influence that development. The macro-scale perspective on history provided by technological determinism can be useful for gaining insights into general patterns of technological development, but if such insights are not substantiated with more detailed accounts of how those patterns are given shape, what remains is a short-sighted interpretation of what are in fact heterogeneous and complex processes.

However, social constructivism also has some limitations that impact its use for analysing the functioning of myths. Most notably, by dismissing the significance of the macro-scale, social constructivists risk becoming myopic, focusing only on the minute details of a process of technological development and then stating the truism that all these details played a part in constructing it. In other words, there would be no clear room for determining whether there are larger-scale underlying forces at work in framing those details at the micro-scale. Furthermore, there is a problem with the fairly common social constructivist practice of seeing technology as a type of text that is ‘written’ by producers and then ‘read’ by consumers. This viewpoint holds that social groups that produce technologies aim to ‘configure the user’ (Woolgar 1991) by converting specific preconceptions about users and their needs into specific technological artefacts, and that users subsequently do not necessarily perceive these the way they were intended. Ultimately, this would imply that technological artefacts have no objective characteristics and that their functioning is always open to radical redefinition.[4] This is problematic because, as media sociologist Ian Hutchby notes, it assumes technology is ‘a *tabula rasa* which is only given meaning and structure through actors’ interpretations and negotiations’, which would ignore that ‘the range of possibilities of interpretation and action is nowhere near as open for either “writers” or “readers” as the technology-as-text metaphor implies’ (Hutchby 2001: 29). When studying myths of idealised communication, a social constructivist approach is difficult to use for adequately considering the notion that both inscriptions and interpretations of technology are informed and constrained by long-held myths, and that such myths thus limit the range of possible developments and uses of communication technologies.

On their own, neither technological determinism nor social constructivism satisfactorily provides the means to find an answer to the question how myths of progress shape communication technologies. Combining the two, however, might work. On the one hand, tracing recurrent myths in a broad historical range of technological developments can tell us a great deal about the ways in which deeply ingrained concepts of hope and purpose are projected upon deterministic, teleological, and similar higher-level forces.[5] On the other hand, myths work because they conceal the failures, the social struggles, and the unreachability of a sublime state; they are extremely adept at setting agendas and letting people forget that they are repeating themselves in proclaiming the gospel of the new. A myth-focused inquiry into the technological condition would shift back and forth from unravelling its complex social construction on the micro-scale, to finding the common strategies that are used to cover up this construction on the macro-scale. It would, in fact, reframe the struggles in the social sphere as the prime realm where myths are upheld. This idea is enforced when we look at a similar argument made by Merritt Roe Smith and Leo Marx in their edited volume *Does technology drive history? The dilemma of technological determinism*. The authors point out as well that, despite the validity of the critique of technological determinism, there is still a poignantly robust feeling in the public imagination that 'technological innovation is a major driving force of contemporary history, if not the primary driving force' (Smith & Marx 1994: xiv). Seeking to explain this apparent contradiction, Smith and Marx defuse the objections made by social constructivism by arguing that these objections merely shift the origin of technology's power to social and cultural spheres, and ultimately do not negate the effectiveness of that power. Thus, they write, technological determinism could well be redefined as referring to 'the human tendency to create the kind of society that invests technologies with enough power to drive history' (ibid.). This observation should not be understood as a call for a return to the study of technology's 'impacts', but as a recognition of how powerful the human need is to think of the flow of technological developments in terms of causality.

In order to pay proper attention to myths of idealised communication in the study of how technology constitutes our environment and vice versa, it is therefore useful to consider a perspective that either transcends or combines the macro- and micro-scale. In the past few decades, several possible theoretical perspectives that do this have emerged, each with their own strengths and weaknesses. There is actor-network theory (ANT), which sees events and actions as constituting heterogeneous networks that link together human and non-human actors. In ANT, as sociologist Bruno Latour has argued, the whole notion of scales actually disappears: there are simply networks with strong and weak links between actors, and studying these networks – or, rather, 'reassembling' them (Latour 2005) – is aimed primarily at revealing how and with what purpose particular actors are connected to each other, and not so much at determining 'order relations' that lead from the micro-scale to the macro-scale or vice versa (Latour 1996: 371). However, while this approach can be advantageous when one wants to get rid of a-priori assumptions about the locus of agency in technological development, it commonly focuses on network interactions over a delineated and continuous period of time, and it can therefore become quite difficult and tedious to diachronically trace recurrent myths that have oriented actors in their mutual configurations. Furthermore, ANT shuns the idea of underlying frameworks; as media sociologist Patrice Flichy has also pointed out, ANT research generally does not consider that technologies are in part the product of 'a collective vision or *IMAGINAIRE*' (Flichy 2007: 4). The ANT approach, then, like technological determinism and social constructivism, would seem to offer only limited opportunities to study the functioning of myths when used exclusively.

Fischer also discusses an alternative approach to technological determinism and social constructivism, one he calls the *symptomatic approach*. In this approach, technology is seen as an expression of deeper forces embedded in social and cultural behaviour, and studying those forces is aimed at uncovering the systemic nature of technology, its *Geist*, as Fischer explains. This approach might provide ample opportunity to study myths, although according to Fischer it risks discarding the diversified ways in which technology can manifest itself, and so could wrongly assume 'that modern technologies form a coherent, consistent whole' (ibid.: 13). Yet, while he is right in arguing that uses and effects of technologies are not homogeneous, his assertion that it is difficult to assess how underlying rationales of technology might work misses the point that describing these rationales can illuminate *expected* uses and *perceived* effects, which do have a powerful converging influence on technology developments. A concrete example might clarify this. Looking at how Fischer presents his argument, it is very ironic that, when he points out the dangers of extrapolating future developments along a path informed and determined by an assumed technology *Geist*, he chooses the proliferation of video games and the computerisation of the home as two prime examples of failed 'grand directions'. At the beginning of the 1990s, when he wrote his book, this might have been a sound observation, but today we know that both the video game and the home computer industries have thrived enormously. Of course, with hindsight it is always easy to ridicule wrong predictions, and I certainly do not want to imply here that every projected future development will come true if we wait long enough, but in this particular instance a macro-scale study of what we could call the video game or computerisation *Geist* could have revealed powerful and higher-level processes at work, which were fuelled by forceful necessary fictions that rendered it very unlikely that they would lose critical momentum any time soon. Blind belief in determinism is problematic indeed, but it is not at all impossible to make educated guesses based on knowledge and understanding of psycho-cultural forces active in the socio-technological realm. So could a reconceptualised notion of Geist help us to study myths of idealised communication in the development of communication technology?

Tracing technology's *Geist*

Social relations are closely bound up with productive forces. In acquiring new productive forces men change their mode of production; and in changing their mode of production, in changing the way of earning their living, they change all their social relations. The hand-mill gives you society with the feudal lord; the steam-mill, society with the industrial capitalist. (Karl Marx 1936: 92)

When critically engaging the various historical notions of the relationship between technology development and its guiding influences, it is quite conceivable that at some point Karl Marx's well-known passage about historical materialism, found in *Das Elend der Philosophie* (translated as *The poverty of philosophy* in 1936), will be encountered. It is famous for the sweeping claim it makes, linking supposedly fixed identities of specific machines, namely those of the hand-mill and the steam-mill, to the direct societal consequences of their mere presence. Although Marx's statement has been revealed as factually erroneous, its rhetorical force remains visible in debates on whether it is insightful or even possible to describe systemic structures in history.[6] Marx was convinced that such a structure could be disclosed; like Hegel, he posited the inevitability of historical progress along a dialectical pattern, a process formed and informed by *Geist*.[7] As we have seen previously, positioning *Geist* as a directive principle in the history of technological development is controversial, to say the least, and not without reason, as it can lead to a reductionist view of why things happen. Yet, equally, dismissing it out of hand for being too deterministic is also not convincing. Postulating the presence of underlying structures is not the same as having a determinist attitude towards how history unfolds. Admittedly, and again taking Marx's notorious statement as an example, a quick glance might indeed give the impression that he describes necessary causal links between modes of production and societal configurations, and, as some scholars suggest, to think that Marx meant otherwise is a mistake.[8] However, as economist Robert Heilbroner suggested in his classic essay on the implications of Marx's claim, *Geist*-engendered relationships in the technological realm can actually be far more complex than what they are when presumed being highly determined; what we should take from presupposing the presence of a technology *Geist* is that it at most positions technology more precisely as 'a mediating factor, both acted upon by and acting on the body of society' (Heilbroner 1994: 63). Outlining and investigating underlying, systemic structures in technology development thus does not necessarily imply that a deterministic or teleological approach is being pursued.[9]

The current challenge is to circumvent the looming criticism of postulating the presence of *Geist*, and to give an overview of recurrent myths in technological developments without falling into the trap of presenting them as if they fixed technology's histories on a single track and left no room for alternative directions or diversions. Such a balanced overview would create a perspective on technology that, on the one hand, takes into account the social, cultural, political, and economic factors that co-construct specific uses of specific technological artefacts, but on the other hand also recognises the psycho-cultural processes that have been present in various guises during technological development, as an undercurrent influencing the ways we think about, develop, promote, use, and value communication technology. This, then, following science and technology scholar Paul N. Edwards, could be called a 'multiscalar approach' to studying technology development, one that tries to describe the 'mutual orientation' of social actors, institutions, and large-scale solutions to persistent 'problems of material production, communication, information, and control' (Edwards 2003: 221-222).[10]

In constructing such a multiscalar approach, I want to draw upon the same insight that informed the theory of *Apparatgeist*, which James E. Katz and Mark Aakhus proposed to define as referring to 'the common set of strategies or principles of reasoning about technology evident in the identifiable, consistent, and generalized patterns of technological advancement throughout history' (Katz & Aakhus 2002: 307). Katz and Aakhus, too, see the inadequateness of constructivist approaches in explaining the persistence of technological myths, and they, too, are wary of taking an alternative but seemingly determinist standpoint: while the authors explicitly incorporate the Hegelian notion of *Geist* in their theory, they are quick to add that although they wish to acknowledge a directive principle in their perspective on technological development, they do not want to include the notion of a teleological goal. They see no given final end; rather, the type of movement they perceive finds its expression both in a sense of continuous becoming – instead of being static and pre determined – and, in a broader historical sense, in that there are 'latent urges' that can suddenly manifest themselves. *Geist*, according to Katz and Aakhus, implies 'incremental change followed by occasional, unpredictable bursts of drastic change in history' (ibid.). So, there is a directive principle present, but its manifestations and conceptualisations are historically contextualised.

Having formulated the main aim with all the aforementioned prerequisites in mind – to create a non-teleological view on the development of communication technologies that acknowledges both incremental and sudden changes oriented by certain directive principles – I propose that a multiscalar approach to understanding myths of idealised communication in technology is derived best from what we could call an evolutionary account of technological history. Before I will explain in more detail why and how I think the evolution metaphor is useful, let me first stress that I recognise that its application can be a precarious undertaking, because, as I already wrote in chapter 1, the understanding of the theory of evolution is anything but uncontested and its foundations are often misused to prove the existence of linear progress. This, of course, is not my intention; I do not want to claim that progress itself is a directive principle in biological or technological evolution. However, I do believe that myths (such as those of progress) play an important role in favouring certain technological solutions over others, similar to how in biological-evolution selection mechanisms enable some organisms to survive while others become extinct. There is no determinism or goal here; neither biological nor technological environments are stable, and processes of generational reproduction regularly mutate through aberrations or are suddenly and violently disrupted by unforeseen events.

Therefore, in my understanding of the evolution of technology, I want to selectively use analogies from the theory of biological evolution, and distinguish the following characteristics: first, a non-teleological view on history that resists the temptation to only concentrate on 'big breakthroughs' or 'brilliant inventors', while still acknowledging the occurrence of chance discoveries or sudden events that quickly surpass paradigmatic boundaries; second, a focus on the importance of environmental factors that can account for the diversity of different uses and appropriations of technology; and finally, a recognition of forceful fantasies, moving myths and seductions of the sublime that work as recurrent selection mechanisms within the evolutionary processes. What follows is a closer look at these characteristics, along with a more detailed explanation of why they fit the evolutionary paradigm.

Non-teleological view on history

A reappearing complexity in writing histories of technologies is that, inevitably, past events are looked at in light of the present. Too often this disposition results in finding – or, rather, constructing – causal links between what has happened and what constitutes our here and now, thereby creating artificial lineages 'characterized by a latent or explicit technological determinism' that suits the needs of whatever interests the historian has or ideologies she believes in (Crary 1990: 8). Within this tendency, the potent idea of discovering the origin of things grabs hold of the historian's imagination, and in the wake of this idea of an origin comes the notion of a *telos*, a final goal that predefines the direction that history had to take. Especially in the realm of technology, where the idea of purposeful design reigns supreme, progress has often been presented as a teleological process: all steps taken along the road form a necessary path between a primitive beginning and a final, ideal destination. In such accounts of progress, origins are mostly found in so-called revolutionary discoveries by individual geniuses, who subsequently have tried at all costs to have their inventions mature into their full-grown selves. As a result of these forms of historiography, artefacts that do not fit the traced trajectory are marginalised as failures and are left for forgotten.

The issue with such teleological histories is not so much that they are linear (although linearity poses its own problems as well), but that they tend to focus solely on successful stories that are lifted out of a continuous and heterogeneous stream of events. In *The evolution of technology*, George Basalla asserts that this notion of discontinuous technological change, which supposes that revolutionary discoveries are miraculously spawned in contextual isolation, clearly needs revision (Basalla 1988: 26-30). Squashing the assumption that humans respond to a universal urge to create technologies out of necessity or to have them suit fundamental needs,[11] Basalla holds that our lifeworld plays host to a diversified abundance of superfluous technological artefacts, all of which can be placed relative to each other in an ongoing evolutionary flow of development. He supports his view by extensively showing that even artefacts 'such as the steam engine, the cotton gin, or the transistor' that 'appear to be excellent candidates for use in supporting the contrary discontinuous explanation', have 'emerged in an evolutionary fashion from their antecedents' (ibid.: 30). This does not imply that those antecedents are all necessary determinants; Basalla rightly adds that within an evolutionary perspective many variants of artefacts can be distinguished in the flow of development, of which '[s]ome are simply inoperable; some are ineffective; and some are effective but have little technological and social influence' (ibid.: 34).

Basalla thus gives us the evolutionary vocabulary to describe and explain continuous processes of technological development without having to resort to accepting necessity, determinism, or progress as instigators and movers of those processes. He also stresses that the continuity argument does not imply that certain inventions are inevitable, or that the paths of development are not subject to alteration by social, economic, cultural, technical, or other forces. Evolutionary processes mutate, or are either accelerated or stopped in their tracks due to coincidences and unforeseen events: '[If] evolutionary change is to occur, then novelty must find a way to assert itself in the midst of the continuous' (Basalla 1988: 63). Novelty is the counterpart of continuity in the general pattern of evolution, generating prosperous deviations, sure dead ends, and with them a myriad of non-determined effects. In technology development, a similar pattern can be distinguished; if there was no occurrence of novelty, Basalla writes, 'strict imitation would be the rule, and every newly made thing would be an exact replica of some existing artefact' (ibid.: 64). New technologies can fundamentally change the status quo, die untimely deaths and certainly have unintended consequences.[12] An evolutionary perspective on technology, when applied in its original Darwinian non-progressive version, is therefore very useful for avoiding teleological notions of development.

Environmental factors

If we are to take an evolutionary perspective on technology and want to consider the role of environmental factors, the next thing that needs to be clarified is to what extent we can push the analogy.[13] In biological evolution, environmental characteristics form the territorial demands imposed on a population to survive, hence acting as a preserver of profitable variations within that population. When the environment changes, either gradually (for instance due to geographical modification) or dramatically (for instance due to meteor impact), the survival demands change as well, disturbing the existing variational balance and shifting it in a new direction. The environment thus 'selects' (albeit not actively) rather than produces variations within an organism's population because these variations are already present, as a result of random natural processes occurring in generational reproduction. The process is random as there is no intentional direction towards a set goal; if there is one thing that we might consider an inherent characteristic of evolution, it is that its course can be typified by constant 'diversification, expansion, and exploration' (Buskes 2006: 426).

In technological development, we can say that there are demands as well, and that their arrangement is highly dependent on the specific environments they are formulated in. We can think of social demands, political demands, cultural demands, and so forth,[14] which all favour different or overlapping types of technology while disfavouring other types. In addition, the environments in which the many variants of technology exist are, just as in biological evolution, multifarious and unstable, and thus we will find major differences in the places, ways, and times in which specific technological artefacts are developed and used – if they are developed or used at all. Media scholar Brian Winston provides a useful way of describing how technological change incorporates these notions of environmental factors. Always grounded in what he designates as the 'social sphere', Winston sees that in the transformation process from idea to prototype to invention to production there are environmental forces at work such as 'supervening social necessities' and 'the "law" of the suppression of radical potential', both determining which ideas, prototypes, inventions, and eventually produced artefacts are suited to survive, and which become redundant (Winston 1998: 3-14).

This view does lead us to recognise that we must be careful not to extend the environmental analogy too radically. The most important difference between biological and technological evolution is that changes in organic variation are the result of random processes, whereas technological variation is, for a large part, actively and paradigmatically produced in response to perceived environmental demands and changes. This does not mean, however, that we can take an instrumental view and assume that environmental factors always induce the right technology to be invented in a rigid problem/solution pattern; moreover, the fact that technology development is encapsulated in a direct feedback loop from heterogeneous knowledge to intentional action, incorporating many levels where all kinds of variation, selection, and responses are to be found, makes it a much more complex process to study.[15] Additionally, in what sociologist Eric Hirsch calls the upholding of 'domestic self-sufficiency', dominant environmental forces (such as the manufacturing industry's inertia and the needs of consumers) co-evolve with technological innovations, strengthening their mutual dependency and legitimising their existence (Hirsch 1998: 159). In other words, it is prudent not to confuse different levels of analysis. By using the environmental analogy I primarily want to stress that technology development is an adaptive process, and as such will continue to uphold a great diversity within its 'population'.

Forceful fantasies and moving myths

Having established that an evolutionary account of technology enables us, to a certain degree, to place technological development in a continuous scheme that, at the same time, possesses no inherent teleology or other forms of linearity because of the many environmental factors that guarantee diversity, we now come to the third and final element of the proposed perspective, which is the focus on recurrent myths that underlie movements in that evolutionary scheme. Why is it that we continue to find very similar projections of and reactions to new communication technologies? Do these projections and reactions somehow prefigure a limited set of conceptions of what constitutes the ideal state of those technologies? An appropriate answer to these questions can be obtained by briefly returning to the evolutionary biology debate in chapter 1 between Daniel C. Dennett and Stephen J. Gould, who differed in opinion regarding the contingency of evolution. Dennett, in response to Gould's opinion that every 'replay of the tape of life' would result in a radically different outcome, argued that the nature of evolution is actually such that it would come up with the same 'Good Tricks' each time (Dennett 1995: 304-308). In other words, according to Dennett the same patterns emerge again and again in evolution, in different times and places and in different guises, because they – somehow – apparently work. This phenomenon is also called 'convergence' – the idea that organisms can differ in their development but will have similar roles when they adapt to comparable ecological niches (Buskes 2006: 427).

In the technological realm it is possible to say that we can see a similar thing happening. Technological artefacts evolve in many ways and are used for multifarious purposes, but not limitlessly so; there are convergent constraints on what does and does not work. In communication technologies, these constraints are what Ian Hutchby calls communicative affordances, the 'social and technological rules' that determine 'the relevant "slots" which delimit the possibilities for actions' (Hutchby 2001: 28). To focus on affordances, Hutchby argues, is to 'accept that technological artefacts both promote certain forms of interaction between participants and constrain the possibilities for other forms of interaction' (ibid.: 32). With Hutchby it is possible to note the important difference with biological evolution, which, as highlighted before, is that constraints of technologies do not randomly originate in the wild; rather, because of the predominantly intentional nature of technological innovation, they are actively, discursively, and materially produced. Hence, notions of 'what works' in communication technology are highly dependent on psychological and culturally cultivated ideas of what communication should be.[16] The power of such ideas has been described earlier in this book: in the search for novelty amidst the continuous stream of technological evolution, we are lured by mental representations of an ideal goal whose expression in turn is rooted in perceptions of both the technological and the communication sublime.

Deeply ingrained mental habits to grasp the 'not-yet-become' of technology are therefore a constituent part of technology evolution, acting as selection processes that both discard and set in motion evolutionary processes. Winston, for instance, sees technologies as the 'utterances' of a certain scientific competence, whereby the transformation from competence to utterance is instigated by 'mysterious mental forces [such as] creativity, intuition, imagination, "the will to think"' (Winston 1998: 3, 5). What is more, Winston appears to hint that these forces automatically manifest themselves according to an internalised and shared set of rules. There is no need for the technologist to be a scientist herself in order to envision what a transformation might look like, Winston writes, because she will, at some level, already have 'absorbed' the 'grammar' of technological competence (ibid.: 5). Basalla as well notes that, for a large part, technological innovation follows the rules of 'the game of invention' which lie 'at the heart of Western technology'; he writes that, next to rational and utilitarian motives, myths such as 'technological dreams, impossible machines, and popular fantasies', which are often coupled with the idea of progress, continuously underlie technologists' decisions and justifications (Basalla 1988: 66-67, 131-132).

Reappearing myths can thus be said to inform and constitute the 'good tricks' of technological evolution. That is, they supply the discursive foundations of 'what works', or rather, of how we perceive 'how it should work'. This in turn sets agendas for further action. Hence, in the development of communication technology we should be able to locate traces of myths of the communication sublime, that over and over again tell us that better communication is possible, if only the technology were perfected and people would learn how to use it effectively. Such myths work like *memes*, the term that evolution theorist Richard Dawkins uses to describe gene-like replicators of information that live in 'the soup of human culture' (Dawkins 1976: 192), or like *topoi*, the term with which Huhtamo defines 'cyclically recurring elements and motives underlying and guiding the development of media culture' that provide '"pre-fabricated" moulds for experience' (Huhtamo 1994).[17] The recurrent nature of myths of the communication sublime implies that they are always potentially there, but that they can be actively deployed or ignored, or as Mosco observes, they can be temporarily alive or dead (Mosco 2004: 29). It also means that, even when myths inform conceptions of technological artefacts that do not fit contemporary paradigms of the environmental constraints in force, they can still tell us something about the reconfigurations within the discourses and material expressions of the communication sublime.[18] Moreover, because there can be many interpretations of how to realise forceful fantasies, we can be certain that large parts of the evolution of technology will be marked by tension and conflict over which direction to take.

The observant reader will have noticed a potential danger in my approach, and it is one that I would like to defuse before continuing with the excavation of myths of the communication sublime that are located in the evolution of media technology. By focusing on the persistent nature of necessary fictions and their power to guide technologists in the direction of what is perceived as 'that which works', the reader might object, there is a risk of reintroducing a teleologically constructed view on technological development. This risk looms large because, as film historian Frank Kessler argued in 'Het idee van vooruitgang in de mediageschiedschrijving' ['The idea of progress in media historiography'] (2002), describing similarities in different technological artefacts (both diachronically and synchronically) raises the problem of assuming the existence of transhistorically preserved 'true' identities of media technologies that ultimately emerge.[19] Such a view, Kessler contends, would render invisible the vastly heterogeneous pragmatic conditions and functioning of those technologies, and would retroactively reconstruct a new teleology (Kessler 2002: 19, 22). To counter the tendency to find a goal-oriented purpose in media development, Kessler proposes to use the concept of the *dispositif*[20] to describe the 'differences that hide behind the apparent identity of separate media', which in turn reveal themselves in the specific constellations of a technology, its modes of addressing its users, and its textual form (ibid.: 21). Using such a perspective, Kessler writes, we would be able to see that, for instance, the dispositif of mobile telephony (despite overlapping characteristics) is very much dissimilar to that of 'traditional telephony', mainly because of differences in mobility and in the nature of sent messages (ibid.: 21-22). In other words, it would be quite inattentive to postulate that mobile telephony is naturally a newer (and hence better) form of older telephonic technology, and shares the same basic identity.

While I wholeheartedly agree that looking for similarities between media, even when their differences are stressed, runs the risk of foregrounding induced media identities, I would contend that this is not what I do. My object of study is not the similarities or identities themselves, but *the desire to see* similarities and identities. By using an evolutionary perspective on the development of communication technology while focusing on myths of ideal communication, I intend to expose teleological thinking rather than practice it myself. Moreover, the principles of evolution allow me to highlight the histories of technological artefacts that in some way illuminate how the discursive formation of communication improvement is constructed, but that at the same time are considered to be 'dead media' within the dominant teleological mindset (Sterling 2006a: 57). This does not mean that the conceptual framework of the dispositif should be dismissed. Without describing dispositifs in too much detail, in what follows I will at irregular intervals stress that these serve well as descriptive concepts with which the structural boundaries of the many branches in the evolution of communication technology can be distinguished.

Myths of ideal communication in media evolution

> *We might say that, inasmuch as 'media' are media of communication, the emergence of a new medium is always the occasion for the shaping of a new community or set of communities, a new equilibrium.* (Lisa Gitelman & Geoffrey B. Pingree 2003: xv)

I am now in the position to give an account of the media environments and myths of ideal communication that have continuously oriented and pervaded the process of media evolution. Although I will travel through media history roughly chronologically, I am working from the premise that many media dispositifs form constellations whose boundaries are fluid, extending beyond rigid sequential structures. Following Zielinski's later work on media archaeology, my intention here is to understand these media constellations as 'spaces of action for constructed attempts to connect what is separated' (Zielinski 2006a: 7). In other words, I see them as embodiments of the numerous material manifestations of myths of ideal communication. Unlike Zielinski, however, whose expeditions are primarily aimed at uncovering the new in the old and not at exposing media *topoi*, I will make my primary analytical cuts at points where the boundaries of media constellations are typically represented as clear-cut, as stable conceptions of media that are aimed at preserving the perception that each new medium constitutes a particular phase in the process of ever-improving communication technology. Each cut will therefore focus on dominant media manifestations, in which myths of ideal communication are located in the imaginative prefiguring, development, marketing, adoption, and use of specific technologies.

For comparative purposes, the analytical cuts are concise and directed at uncovering hopeful expectations and predictions. This means that I will not describe in detail the many anxieties and fears that also accompanied the advent of new media, although at times I will show that such negative reactions themselves have functioned as a backdrop against which newer (and thus 'even better') technologies were being positioned. The focus lies on five media constellations, respectively: the telegraphic, the telephonic, the radiographic, the televisual, and the Internet constellation. In each constellation I will describe the various actors, technologies, motives, and myths of ideal communication that played their parts in orienting media evolution, and show that these myths moved dynamically across constellations. As we will see, recurrent discursive elements in the necessary fictions of communication improvement consist of *topoi* of world peace, ultimate understanding, cultural and social unification, unlimited progress, a return to a pre-Babel existence, the sharing of a universal language, a relief of all anxiety, the realisation of utopia, and so forth; these are all very much recognisable from the analysis of notions of progress and idealised ideas of communication in chapters 1 and 2. Despite the fact that these hopes and dreams have all been denied by reality, they keep resurfacing and keep setting agendas for further development.

The telegraphic constellation: building the network paradigm

In the semantic field of communication, locating the starting point of mediated communication is arbitrary. From an evolutionary perspective, there can actually be no single beginning at all: there are always antecedents. We could suggest that the lineage of communication technology can be traced back to the time when people started to send messengers (human or animal) across distances farther than a human voice could carry. Or, taking the need for a physical carrier out of the equation, to the moment that beacon fires, clouds of smoke, flags, or other materials were used for the very first time to instantaneously transmit signals to distant observers. Whatever the means, though, the goal or desire was always to bridge space and time. It is the most powerful myth of communication: no distance should be an obstacle for communication to take place.

Up until the beginning of the seventeenth century, the distance over which optical communication systems could reasonably be expected to function was limited by what a (sharp) human eye could see; after the arrival of Galileo's telescope, this limit increased to about 30 kilometres. Around the same time, theories were being explored around the theme of instant and unlimited communication at a distance which involved 'magnetic forces' to construct a signalling system. Inspired by the writings of Giambattista della Porta, the Italian Famianus Strada, for instance, wrote in 1617 that by magnetising two compass needles it would be possible to have them move in rapport, thus establishing a channel over which one could 'express all the sentiments of your mind' to a 'friend who is far distant' (Strada 1617 cited in Holzmann & Pehrson 1994: 41). Although such musings were primarily imaginative concoctions that arguably belong to the realm of 'non-inventions' (ibid.: 40), they do vividly convey how ideas of idealised communication tell of eliminating distance, and of opening up a media space where there can be an almost telepathic and uninhibited exchange of thoughts or feelings. Moreover, writes Zielinksi, such fantastical narrative constellations always 'signify something' about the twilight realm that is the world of media: 'the relationship between what is imagined and what in fact exists, between (mere) fantasy and (actual) reality is fluid, unstable' (Zielinski 2006b: 29).

While media conceptualisations like those of Della Porta and Strada characterised the seventeenth- and eighteenth-century socio-cultural imagination of what communication could be like if it were freed from its carrier, their suggested use of long-distance communication technologies was primarily of a romantic nature. This perception of communication did not correspond with the strategic functionality that had typified the use of optical devices up to that time. In Winston's terminology, for a long time there were no supervening social necessities that asked for a functional transformation of the ad-hoc and utilitarian make-up of existing long-distance communication systems. Necessities, however, changed significantly at the end of the eighteenth century in France, stirred by the Revolution. At the heart of these necessities lay the desire for increased speed of communication: the French Enlightenment ideal to implement a unified and national identity called for a unification of language as well as for the means to quickly organise and consolidate that unity. This project was perfectly in tune with the creation of the semaphore telegraph by Claude Chappe. With help from his brother Ignace, who had been elected to the French Legislative Assembly and had arranged for Claude to have his idea approved and subsidised, Chappe set out to devise 'a simple method for rapidly communicating over great distances, anything that could be the subject of a correspondence' (Chappe 1792 cited in Holzmann & Pehrson 1994: 56). While emphasising the gain in speed that the new communications system would bring, Chappe also stressed the advantages of near simultaneous communication: 'It offers a reliable way of establishing a correspondence by which the legislative branch of the government could send its orders to our frontiers, and receive a response from there while still in session' (ibid.). A longing and need for real-time communication is clearly visible in Chappe's line of thinking, propagated by both military and scientific motives. His system, which consisted of towers that were placed at regular intervals and that supported a set of rotating arms with which hundreds of symbols could be encoded, was subsequently built and appropriated to explicitly support the Republic's cause: use swift and standardised communication to defend the borders and to write a shared 'Republican language' onto the mental landscape of the French people (Flichy 1995: 14).

Even though in regular use the Chappe telegraph system was mostly assigned a military and political function, and ordinary citizens had very little access (Headrick 2000: 197-198), it contributed greatly to the popular idea of the possibility of instant communication, which characterised the European utopianism of that time. The notion that new technologies could play an important role in realising that idea was widespread; Chappe was most certainly not the only one occupied with enhancing technological devices that could quickly transmit signals. Many amateur inventors in many countries did the same, either in fiction or with concrete experimentation (Lardner & Bright 1867: 39-44). Out of these experimentations, in the first quarter of the nineteenth century, came another possibility of speeding up communications. Magnetism and dynamic electricity re-emerged as means to transmit information over a distance, but this time in scientifically much more viable ways than described by Della Porta and Strada. A new paradigm for swift and simultaneous communication appeared at the horizon in the form of the electric telegraph, which, according to communications theorist James W. Carey, 'permitted for the first time the effective separation of communication from transportation' (Carey 1989: 203). Although this observation is, in light of Chappe's semaphore telegraph, not quite true, the instantaneous character of electricity and electrical communication technologies did spark a feverish belief in the seemingly limitless possibilities of 'communication without embodiment, contact achieved by the sharing of spiritual (electrical) fluids' (Peters 1999: 139). The rhetoric of the 'electrical sublime' was filled with notions of communication utopias; as Carolyn Marvin writes, '[t]he power of new electrical technologies to effect effortless intimacy between friends was perhaps the commonest of all prophetic themes about communication' (Marvin 1988: 154). The advantages to be gained by using electric instead of optical telegraphy were stated quite clearly in familiar obstacle-removing and thus communication-improving language: less costly to build, greatly enhanced speed, and no more 'liability to interruption from misty weather' (Lardner & Bright 1867: 40).

It did not take long for all kinds of variations of electromagnetic telegraph systems to evolve from the newly fertilised ground of scientific competence; there were systems that used galvanometers, ones that worked with electro-chemical parts, and ones that were powered by electromagnets (Highton 1852; Sabine 1867). Emphasising the idea of 'good tricks' in media evolution, these variations sprang from a diverse group of inventors, all of which were, to a certain degree, aware of the range of possibilities of electricity. So, while history books often name Samuel F.B. Morse as the sole inventor of the electric telegraph, from an evolutionary and historically correct point of view he was but one of many. What does make Morse, as well as his British contemporaries William Cooke and Charles Wheatstone, stand out from other experimenters, is that in the 1840s they were the first to successfully commercially exploit a fully functional telegraph system. Whereas the optical telegraph was invariably under exclusive state control and the first electric telegraphs barely left the laboratory, Cooke and Wheatstone in England and Morse (and his forgotten assistant Alfred Vail) in the United States made their telegraph systems readily available for paying customers, and sought to quickly expand their networks. In both countries, the relation with railway companies was strong; not only did the electric telegraph provide a solution to the supervening social necessity of railway safety, but both technologies also complemented each other in conspicuously portraying ongoing technological progress. Especially in the Unites States, where frontier mythology already drew heavily from railroad symbology (Schivelbusch 1986), electric telegraphy was hailed as 'the nerve of international life, transmitting knowledge of events, removing causes of misunderstanding, and promoting peace and harmony throughout the world' (Anonymous 1868 cited in Standage 1998: 91).

The perceived mystical characteristics of instantaneous communication through electric telegraphy influenced popular imagination so strongly that they seeped into the preferred metaphors that spoke of making contact with any distant body, including the dead. This is poignantly visible, for instance, in late nineteenth-century Spiritualism. As Jeffrey Sconce writes, for the Spiritualist movement, 'the bodiless communication of telegraphy heralded the existence of a land without material substance, an always unseen origin point of transmission for disembodied souls in an electromagnetic utopia' (Sconce 2000: 57). Sconce demonstrates how idealised ideas of communication conjure up visions of an ultimate togetherness, facilitated by a technology that in practice merely established Morse-encoded point-to-point contacts, but in the social imagination carried with it the unifying promises of hope, purpose, progress, and the new. Such expectant ideas were not confined to semi-religious circles, however; in regular everyday use the electric telegraph also strengthened the notion that it caused a revolution in the enhancement of social organisation. News poured in more quickly than before and from ever-more remote locations, thus reinforcing the notion of 'internationality' (Mattelart 2000: 17). Sophisticated 'telegraph exchanges' in the 1860s made the switching of private lines possible and facilitated real-time telegraphic conversations between banks or lawyers (Standage 1998: 172-173). Businesses discovered the electric telegraph as a way to track the latest developments of the stock market on an equal basis, reducing the chances of monopolies (Du Boff 1984: 571-572). Travellers informed their family and friends of their journeys. People declared their love to each other. The motives for using the telegraph were, in the end, very similar; they revolved around speeding up communication and conquering space, for whatever reasons. As long as the obstacles of distance were seemingly removed and people were connected, the technology was perceived as answering the long-held call for improvement in communication.[21]

Looking back, telegraphy, both optical and electric, established the basic characteristics of telecommunications technologies: it constituted a growing and (eventually) permanent network, operated by a specialised technical body, that facilitated fast communication, encoded in a universal language (Flichy 1995: 31-32). In this sense, many current telecommunications systems are 'telegraphic' in some respect, and we can point to several nineteenth-century telegraphic equivalents of what are generally perceived as more modern-day technologies.[22] It is furthermore obvious that, despite all the optimistic outlooks, telegraphy did not bring world peace; perhaps on the contrary, it proved to be a very efficient tool for those in power to coordinate war efforts and to rule colonies.[23] Telegraphy also did not spread to the masses as quickly as many believed or hoped it would, nor did it have the inherent capacity to always bring people closer together in blissful harmony. Because of state monopolies, the requirement of expert knowledge to operate the technology, and the notion that telegraphic messages should primarily serve non-frivolous purposes, in the early years it was mostly governments, big companies, businessmen, and rich individuals that used and had immediate access to the telegraphic service. Over the course of several decades this had largely changed, as many post offices had started to offer telegraphic services to the general public, but competition from the rising telephone system at the beginning of the twentieth century meant that telegraphy remained a technology that was not used on a day-to-day basis by ordinary people. Today, telegraphy has become almost extinct, and has only survived as a nostalgic and often expensive way to communicate.[24] Its myths, however, remained very much alive, and quickly resurfaced during the development of telephony.

The telephonic constellation: getting personal

While electric telegraphy gained momentum in the second half of the nineteenth century, experimentation with electricity and magnetism continued. Because the true nature of the power of these natural forces was still very much unknown – which made them daunting and tempting at the same time – there were many wonderful dreams of what their application could be. It is not surprising to see that among them was the idea to use electricity to transmit speech, a notion clearly informed by the age-old desire to communicate from afar; as telephony's first historian Herbert Casson determinedly writes, the invention of electric speech 'was the result of a persistent and deliberate search' (Casson 1910: 13). Just as optical mechanisms had been used for some time to artificially extend the eye, techniques like yelling or technologies such as speaking tubes or string telephones[25] had been well-known ways to extend the voice; unsurprisingly, electricity and the success of the electric telegraph caused new and heightened levels of expectation that these extensions could be improved. The environment to which the idea of transmitting speech by electricity sought to adapt itself was not entirely favourable, though. Many experts in the field of telegraphy, for instance, thought it was impossible and rather sought to improve the transmitting capacity of telegraph wires instead. The history of the invention of the telephone is therefore one that is often charged with anecdotes that tell of persistence in the face of great hardship and of accidental discoveries that proved to be fundamental breakthroughs. The reality is that telephonic devices slowly evolved through technological tinkering by many experimenters, and from the 1850s onwards appeared in many ways and forms.

The person who is credited as the first to have constructed a working telephonic system is the Italian inventor Antonio Meucci, who is believed to have done so as early as 1849 (Catania 2003: 111). His work, however, did not gain much public nor scientific attention, mostly because of his financial troubles.[26] Then there is Charles Bourseul, a Belgian-born French telegraphist and inventor, who is often mentioned as being among the first to describe a feasible mechanism for converting sound waves into electrical signals (Du Moncel 1879: 13-14). Although he himself did not work on an actual machine, Bourseul's 1854 proposal was read by numerous scientists, propagating the electric speech meme like pollen grains in the wind. One of the responsive receptors was professor of physics Johann Philipp Reis, who built several prototypes of what he called the 'Telephon' during 1858 to 1864 (Aitken 1939: 17-18). His inventions were not quite suitable for electrically transmitting speech,[27] but they did lend scientific credence to the actual possibility of doing so. It is also highly conceivable that Elisha Gray, Alexander Graham Bell, and Thomas Edison, the inventor-entrepreneurs usually associated with the 'true' invention of the telephone, all knew of Reis' devices, and had started to build upon his or similar findings (Coe 1995). We could therefore say that, by the 1870s, the coming of the telephone was in the air, even though there was no particular demand from the general public, let alone from the growing telegraph business. Applying newly gathered electromagnetic knowledge to build a telephonic device was an ambiguous quest, one in which experimenters had to divide their focus between advancing science on the one hand, and meeting the needs of the growing telegraphic market on the other hand by improving already existing telegraphic solutions.

It is worth pointing out here the motives of Gray and Bell, the two adversaries whose caveat and patent application for a telephonic device were both filed on 14 February 1876, separated by only a few hours.[28] Gray, who was an expert in telegraphic technology and worked most of his time on perfecting the 'harmonic telegraph' (with which multiple messages could be sent through the same wire), had discovered a way to transmit speech, but he was rather reserved about what its profitable application could be. Bell, on the other hand, while pressed by his sponsors to also work on multiplex telegraphy, did have a clear idea of the advantages of telephony, and secretly dreamt of and worked on improving the telegraph by replacing it with a 'voice-carrying machine' (Casson 1910: 25; Watson 1926: 62-63). So, while the invention of the telephone is often portrayed as a purely accidental event that occurred while Gray and Bell were working on something else, this apparent contingency is misplaced; both men were very much aware of what they were doing, and it was Bell who was most captivated by the dream of improving communication through technological means. A year after Bell (and his ever-faithful assistant Watson) had managed to capitalise on his patent and construct a working telephone, he advertised its advantages as such:

> 1) No skilled operator is required, but direct communication may be had by speech without the intervention of a third person.
>
> 2) The communication is much more rapid, the average number of words transmitted in a minute by the Morse sounder being from fifteen to twenty, by telephone from one to two hundred.

3) No expense is required, either for its operation or repair. It needs no battery and has no complicated machinery. It is unsurpassed for economy and simplicity. (Bell 1877 cited in Casson 1910: 53)

Bell here clearly appealed to the susceptible sentiment for unifying closeness through improved communication. The emphasis on the ease and directness of telephonic communication is evident, as is the positioning of the new medium as being superior to older technology. The impact that this act of idealising communication has on the human psyche is not to be underestimated; as Colin Cherry writes, by offering the availability of a true conversation – which according to Cherry is 'an essential human relationship' – Bell's telephone promised to help 'resolve uncertainties, doubts, or anxieties and give greater *assurance*' (Cherry 1971: 49, emphasis in original). This does not mean that social bonding was the logical object of its primary use; the telephone first became a business tool once it had overcome the inevitable growing pains that are typical of all fledgling network technologies, and even when it had entered the homes of tens of thousands of ordinary people, it was for a long time thought of as for serious use only (Fischer 1992: 66). Nevertheless, the telephone's inherent capacity to bridge space virtually instantaneously and connect people in a natural conversation was strategically used in advertisements for strengthening the idea that communication had indeed improved.[29] Whether it was for business, groceries, or social cohesion, a telephone call signalled communicative progress. This was achieved not so much by inherently reducing misunderstanding, but by offering a sense of what I would call 'assurance within reach': the conviction that the ever-expanding telephone network would be at hand to come to the rescue, quicker than ever before, in case someone wanted to be heard or get informed. Spoken dialogue from afar was now finally possible; maybe not as ideal as in a Socratic face-to-face configuration,[30] but nevertheless providing an improvement over telegraphic conversations.

It should be noted that this affordance of connecting people in dialogue was not yet visible in Bell's early demonstrations of his device. Although his intention had been to make two-way conversation possible over long distances, his first experimental telephones in 1876 were technically limited to transmission in one way only (Aronson 1977: 20). This forced Bell and Watson, in their probing of supervening social necessities, to give what we now would call broadcasting performances, in which they sent music, news, and dramatic plays over the wires. They continued to stage these entertaining demonstrations even after they had perfected two-way communication, mainly because it earned them a lot of easy money (ibid.: 20-23). The upshot of this was that in the popular mind an additional dispositif of telephony was envisioned, one in which its use became associated with the ability to call in on operas or to subscribe to regularly scheduled news or entertainment programmes – services that were indeed available in cities such as London and Budapest from the early 1890s until the 1920s (Marvin 1988: 209-216; Briggs 1977).[31] Quicker than the newspaper and more readily available than a private telegraph, such telephonic systems prefigured the idea of radio broadcasting, carrying with it a sense of communally connecting to others.

Eventually, the conversational model of the telephone prevailed. With further improvements implemented (most notably the exchange switchboard and Pupin loading coils, which greatly increased the connectivity and dimension of the network) and steady expansions taking place in many countries,[32] the telephone system evolved into the universal service that Bell and his company's president Theodore Newton Vail had envisaged, connecting even the most rural locations to the growing global network (Pool et al. 1977: 130-132). Advertisements for the service evolved with it, strengthening myths of improved communication that held that universality in connectedness is a 'good trick'. Until the mid-1920s, telephony was praised most for its practicality in situations of housekeeping, emergencies, and business deals, thus stressing the improvement of instrumental communication as the telephone's 'natural' purpose (Fischer 1992: 65-69). It was only after telephone executives had decided that social use of the telephone, which had soared ever since the beginning of the twentieth century, was not trivial and actually represented a valuable market, that ads began to focus on presenting non-instrumental reasons for having a telephone. The underlying gist of the message of these telephonic ads, however, remained the same: connect to the growing network, enhance your communication possibilities, make sure that you can reach out or can be reached when necessary, for whatever reason. As explicitly visible in the work of Casson, the idea that new technologies could further promote harmony was strongly present; at the end of his book on the history of the telephone, he writes: 'Who could have foreseen what the telephone bells have done to ring out the old ways and to ring in the new; to ring out delay and isolation and to ring in the efficiency and the friendliness of a truly united people?' (Casson 1910: 298).[33] Much more than telegraphy had done, telephony and its myths signified a slice of modernity available to the masses, promising to transform its users into earthbound angels that could be in electrical contact with anyone else.

Today, it would strike us as odd to think of the fixed telephone as a device that provides an extraordinary, transcendental experience of communication. Because of its ubiquity, ease of use, and capacity to connect people in dialogue, it has become a mundane, almost taken-for-granted way of connecting to other people; so much so, in fact, that it is easily forgotten, for instance, that the global infrastructure of this 'old' technology made possible – and still supports – large portions of the infrastructure of the Internet, that other global telecommunications system that is surrounded by myths of improved communication. Practically gone are the claims that international telephony would bring about world peace by making us all equals, aided by the development of a 'common language, or common understanding of languages' (Pool et al. 1977: 128-129). Telephone users did not suddenly want to reach unknown others purely because there was a potential connection; instead, the telephone became a great tool for maintaining existent networks, even if these had expanded greatly under the influence of, for instance, urbanisation. Misunderstandings also did not subside; moreover, while people can enjoy the bliss of talking to each other while physically separated, they will at times also feel the 'frustrating immateriality of telephonic communication', which does not annihilate space but stresses it, and which prevents the conversation from ever attaining the argumentative closeness of a face-to-face dialogue (Young 2003: 230). However, myths of improved communication remained present in the evolution of media technologies, especially so with the advent of radio communication.

The radiographic constellation: magic in the air

In the second half of the nineteenth century, surfing on the waves of triumphantly hailed achievements in harnessing magnetism and electricity for the benefit of improving living conditions, there were two separate but soon to be linked problems that inspired physicists, experimenters, and inventors alike to further investigate those puzzling electromagnetic powers. The first challenge, which had already occupied Newton when he studied the characteristics of light, was to understand how natural forces actually propagated through space, and how they connected or affected distant physical objects without touching them. The answer, so it seemed, lay in postulating the existence of the *æther*, which was believed to merge together and make possible the forces of light and electromagnetism.[34] Even though the plausibility of the theory of the æther was questioned and disproved at the turn of the twentieth century,[35] its mythical attractiveness was strong enough to spur metaphysical speculations about universal communication. Conceived of as an invisible, omnipresent 'fluid' or force field that could carry waves and vibrations across endless distances, the æther promised telepathic-like connections to anyone, anywhere (Czitrom 1982: 65).

Apart from these scientific agendas that were set by myths of unlimited communication, there was a second incentive for engaging in experiments with the electromagnetic spectrum, one that was informed by the very pragmatic desire to improve electric telegraphy by removing its dependency on poles and wires. Already at its inception it was clear that the electric telegraph network could only grow as much as its physical linkages would allow, and that its connective potential would increase almost infinitely if material constraints were to be taken away. In the years following the successful commercial exploitation of telegraphy, therefore, attempts were made to establish reliable telegraphic contact by natural conduction, either through water or the air. Just as Bell had deliberately sought to improve telegraphy by making the transmission of human speech possible, so did the prospect of improving telegraphy yet again prompt further experimentation. It was a matter of time before the memes of this technological project would meet up with those of research on the æther. Considering its suspected omnipresence and unlimited power to carry signals to anywhere in the universe, it is no wonder that – especially in popular imagination – the æther had acquired a mystical overtone that held within it all the promises of instantaneous and seamless communication. If the æther could be utilised for transferring waves of information, it was thought, it would enable people to be in perfect rapport, without obstacles.

At this junction between scientific research and the age-old drive to improve existing communication means, we find a constellation of well-known and lesser-known figures, all of which have in various ways attributed to the development of wireless telegraphy, and later of wireless telephony or radio.[36] On the scientific side, it was James Clerk Maxwell, a Scottish mathematician and physicist, who in 1864 unified the knowledge of his day on the behaviour of electricity and magnetism into a single set of equations, forming a 'crucial element in the ground of scientific competence' that lay at the base of 'all post-telegraph telecommunications technologies' (Winston 1998: 33). It was not until 1887 that German physicist Heinrich Hertz managed to demonstrate the validity of Maxwell's theory, after having improved the theory to make it account for his own discoveries on the nature of electromagnetic waves. Hertz was not alone in studying these waves; three years after Hertz's experiments, French inventor Edouard Branly built a basic device for detecting the alternating current of radio signals that in later years proved to be a very useful tool for building radio sets. All these developments took place in laboratory conditions, however, and were not intended to be used in the creation of new and commercially viable technology systems. Notice, however, how at this point the imaginative mindset, occupied by the desire to capitalise on the advent of new technologies, revived old promises of communication improvement by emphasising the latent possibilities of those new radio technologies. In 1892, Sir William Crookes[37] published a highly influential article entitled 'Some possibilities of electricity', in which he foresaw 'the bewildering possibility of telegraphy without wires, posts, cables, or any of our present costly appliances' with which 'two friends living within the radius of sensibility of their receiving instruments [...] could thus communicate as long and as often as they pleased' (Crookes 1892: 174-175). This, he claimed, was 'no mere dream of a visionary philosopher' because

> [a]ll the requisites needed to bring it within the grasp of daily life are well within the possibilities of discovery, and are so reasonable and so clearly in the path of researches which are now being actively prosecuted in every capital of Europe that we may any day expect to hear that they have emerged from the realms of speculation into those of sober fact. (Ibid.: 176)

Crookes, then, by using his knowledge of existing scientific competence, and, critically, by envisaging wireless telegraphy as a signalling technology, set the agenda, and redirected the first generation of radio inventors in a teleological manner to the familiar fertile breeding grounds that lie at the crossroads between technological development and communication improvement. To be sure, Crookes was not the only one at that time to ponder the future of wireless communication, but his article provides a rich illustration of how popular imagination of communication progress was stirred in a way that was entirely in tune with late nineteenth-century technological triumphalism.

A few years after Crookes' speculative vision, several inventors (Alexandre Popov in Russia, Henry Jackson in England, and Guglielmo Marconi in Italy) concurrently managed to arrive at working wireless telegraphic devices, using their knowledge of Hertzian waves and Sir Oliver Lodge's improvements of the Branly receiver.[38] It was by chance, perhaps, that Marconi was the first to do so (he conducted his experiments in 1894 and 1895, whereas Popov and Jackson tested their systems in 1896), but he was certainly the one most obsessed with increasing the distance over which he could send and receive wireless signals,[39] and with finding a social use for his device (Aitken 1976: 191). Both these obstacles were overcome when Marconi observed that 'the higher the pole the greater would be the mileage' (Dunlap 1937: 48-49), and subsequently, with help from his English mother, set up his own business and offered his wireless telegraphy services to the British navy. Especially in maritime environments his technology was a success, despite its perceived drawback that it could not provide the shielded point-to-point connections that Marconi originally was after. The inherent affordances of wireless telegraphy would become painfully evident with the *Titanic* disaster in 1912, which marked a dubious high point of what Marconi's invention had brought the world; the possibility of ubiquitous contact could save people from death, take away anxiety for people at home, but also create a cacophony of radio noise as many amateurs filled the æther with their messages. The development of communication technologies, clearly, knows many rocky roads and setbacks, but in myths of improved communication, these are often forgotten or made invisible.

All the while, the evolution of communications technology continued. Propelled by subsequent work by English electrical engineer John Ambrose Fleming and American inventors Reginald Fessenden and Lee de Forest, wireless telegraphy at the beginning of the twentieth century had evolved a branch into the direction of transmitting wireless speech, emulating the two-step process from electric telegraphy to wired telephony. Myths of improved communication received a new boost, as components to build a radio became readily available in the 1910s, and wireless technology was embraced by enthusiastic amateurs who were lured by reinvigorated promises of instantaneous free communication. It was more than just scientists and news reporters that hailed the æther as the all-connecting medium and wireless technology as its nervous system; now, anyone with a radio set could enter an all-accessible communication space, where radio would spread 'mutual understanding to all sections of the country, unifying our thoughts, ideals, and purposes, making us a strong and well-knit people' (Frost 1922: 18). The wonder of instantaneously sending and receiving at a distance via invisible and all-penetrating radio waves fuelled the desire to cross immense distances. To many radio enthusiasts, it did not matter what they were listening to; the sheer possibility of reaching out and touching someone was enough to spend countless nights 'DXing', searching the radio spectrum for distant voices.[40] The prospect of stumbling upon alien signals sparked a true Mars-mania, for 'contacting Mars would be the high priority of the new radio future and the ultimate "catch" of DX fishing' (Sconce 2000: 102).

The properties of radio seemed to perfectly encapsulate the recurrent dream of universal and direct communication, which had already been intensified by the improved point-to-point communication of the telegraph and telephone. Moreover, whereas those latter two technologies predominantly provided individual mediated closeness, radio added a new, more public communicative dimension. It was not just telegraphy or telephony without wires; everyone with a receiver could tune in and feel connected to a virtual community. Because of the messianic character of live broadcasting, the popular idea took hold that radio could be a tool to establish social cohesion and world peace, bringing direct democracy and global harmony to the people. Universal enlightenment, as nineteenth century's 'ultimate visionary crank' Nikola Tesla saw it, 'could be incarnated in the all-pervading waves of the wireless'; for him, there finally was a technical solution to all the problems of communication (Davis 1998: 69, 72). Early forecasts of and experiences with radio broadcasting, in short, expanded the understanding of improved communication to include the possibility of ubiquitous speech, and with it the power to gather people into a single, harmonious, even utopian society.

But, as with telegraphy and telephony, such idealised ideas of communication were never truly realised; they primarily remained strategically used discursive arguments to give radio technology its momentum. During the First World War and in the following years, a new dispositif of radio emerged and gained in dominance. While the dissemination of 'we-ness' at first had been the result of ad-hoc networking by amateurs, in many countries around the world the power of instantaneously reaching large audiences became the monopoly of governments, the military, and media corporations, forcing amateurs into a small segment of the radio spectrum. Two-way radio communication chiefly made way for one-way entertainment, lectures, and news programmes, causing a disgruntled Bertolt Brecht to remark that 'the radio should step out of the supply business and organize its listeners as suppliers' (Brecht 1964 [1932]: 52). Although what remained in this new dispositif was radio's gripping potency to deliver the feeling of absent presence, to capture the 'there and now' and bring it to the here and now, the utopian image of radio communication slowly dwindled. It gave way for domesticated electronic kinship, mingled with an acute awareness of distances, reaffirmed anonymity, and isolation (Sconce 2000: 62). Today, radio is still a very powerful and widely used medium – especially in areas and countries where the telecommunications infrastructure is not strong – but its aura of unhindered communication potential has subsided. What did not subside, however, is the recurrent process of the rise and fall of myths of improved communication, as we will see with the development of television.

The televisual constellation: virtual presence

In his voluminous book *Le vingtième siècle [The twentieth century]* (1883), the first of a trilogy on late nineteenth-century visions of the future, French novelist and illustrator Albert Robida explored the wonders of electricity and technology to come, extrapolating many of the vibrant myths of communication of his time to their supposedly logical outcomes. Writing in a period in which there seemed to be no end to great discoveries by brilliant inventors, Robida gives us a rich account of contemporary technological fantasies at work, describing what daily life in the futuristic 1950s might look like. Among the abundant private 'aerocars', telephones – which he predominantly frames within the radio-like dispositif – and 'photo-paintings', he also describes the 'telephonoscope', with which audio as well as visual signals could be transmitted. Considering the name he gave to this new medium, it is very likely that Robida had found inspiration in *Punch's Almanack*, a popular magazine that in 1878 had published an illustration by George Du Maurier, depicting two parents audiovisually conversing with their daughter through the use of 'Edison's telephonoscope'. For Robida, as film and television historian William Uricchio has noted, the new medium was to be a communication tool that 'could bring distant entertainment into the living room, that could serve as a means of surveillance, and that could serve the mission of "la suppression de l'absence" by facilitating real-time face-to-face communication over vast distances' (Uricchio 2004: 129).

The telephonoscope did not exist at the end of the nineteenth century, but audiovisual media were fully imagined and expected to be invented soon. Not only Robida and those working at *Punch's Almanack*, but many other people as well were captured by the popular idea of using electricity to see farther, and they effortlessly fused existing telecommunications technologies together into new imaginative devices. These visions were fuelled by the apparent ongoing redemption of the dream of perfected communication; the transmission of still pictures via the facsimile telegraph had been a reality since the 1850s, and with the arrival of the telephone and the cinema the desire to come closer to communication utopia by not only extending the ear but also the eye gained in force. Thus,

> [t]he discourse of technical knowledge and the discourse of popular utopian fantasy produced a discourse symbiosis, in which the semantics of the old utopia and the numerous projects for its technical realization, or technical projects for picture transmission in general, were speculatively combined with each other; 'seeing by electricity', 'distant electric vision' was viewed as (another) technical miracle, whose realization seemed to have moved closer in time. (Elsner, Müller & Spangenberg 1994: 110-111)

The histories of audiovisual technology therefore contain a continuous presence of a familiar reasoning in the evolution of communication technology, through what could be called a discursive remediation: with each new generation of media, whether it survives or is stillborn, the promise that communicative utopia is near is resuscitated and fostered within technological, scientific, and popular discourse.

In the genealogy of recording, transmitting, and displaying moving images, we can distinguish a loosely connected and two-tracked pattern of evolution that is similar to those of the previous media constellations that have been discussed. On one track, we find the ongoing scientific research into the characteristics of light, magnetism, and electricity, in which discoveries were made that later proved to be suitable for early televisual experiments, but which were not necessarily the result of intentional attempts to construct such a thing as a 'television'. Among these discoveries, all made in the last quarter of the nineteenth century, were the photoemissive qualities of selenium, electrons (the parts of the sub-atomic world that enabled this photoelectric emission to occur), and the cathode-ray tube, with which those electrons could be made visible to the eye. Instrumental in this phase as well was the work of Paul Nipkow, who from 1883 to 1885 devised a way to mechanically scan an image with a spiral-perforated disk, and electronically send its properties to another rotating disk with which the image could be reconstructed. Nipkow himself did not pursue the possibilities of his 'Elektrisches Teleskop' any further, but, as was the case with the telephone and wireless signalling, the laboratory findings would eventually be appropriated by others.

This brings us to the second track along which television development evolved, the one traversed by the visionaries and tinkerers who dreamt of improving existing means of communication and were trying to make a business out of it. Two of those inventor-entrepreneurs, Charles Jenkins in the United States and John Logie Baird in Britain, independently managed to construct working prototypes of electromechanical television in 1925, using the Nipkow disk design.[41] Of the two, Baird was the most outspoken in emphasising the linear progress he thought he had made in telecommunications. Publically advertising his 'televisor' in 1926 with the theme 'Television For All', he referred to the forecast made in 1876 that '"seeing by telephone" would follow naturally from "hearing by telephone"', and said that now 'the claim had been justified' (Baird 1926 cited in Briggs 1977: 47).

Clearly, Baird's rhetoric is filled with the typical notion that mediated communication is improved by simulating face-to-face communication as accurately as possible. By describing it as a natural process that mediated hearing would be followed by mediated seeing, he uses the familiar strategic discursive pattern of myths of progress. The storyline is that it was inevitable that television would come to be, as it provided a better opportunity to be present where we are not, to get to know things we could not know before, to finally communicate in the way we are 'naturally' used to. For Baird and his enthusiastic contemporaries, new technologies like television promised to let us take yet another seemingly logical step along the path towards the fulfilment of our communicative potential.

As with earlier media constellations, this was a widely held belief. Stressing the presence of 'good tricks' in technology evolution, Jenkins and Baird were not the only ones taking on the quest to commercially exploit the desire and technological possibility for the transmission of audiovisual material, nor did their technical approach represent the only possible way to do so. Concurrently with the attempts to find an electromechanical solution, there were experiments with all-electronic methods to capture and display moving images. Proposals for such electronic systems had been made in 1908 by the English physicist Campbell Swinton, while, unbeknownst to Swinton, a year before that the Russian engineer Boris Rosing had actually already constructed an electronic receiver (Abramson 1987: 26). The ground of scientific competence for making electronic television work was thus well established in the 1920s. In 1923, the Russian Vladimir Zworykin, one of Rosing's former students who had fled to the United States, showed his employers at Westinghouse Electric that electronic television was feasible, and he improved the system at the Radio Corporation of America (RCA) in 1933 by developing the iconoscope (Zielinski 1999: 140-142, 168). During the 1930s, direct comparisons between electromechanical and all-electronic solutions showed that the former had become a non-viable evolutionary branch, delivering Zworykin the title of inventor of television. Again, though, it must be noted, there were many more scientists and technical enthusiasts who published or worked on electrical television; the idea had for a long time been in the air, sustaining the myths of communication and technology that had accompanied many of television's antecedents. Without Zworykin, televisual systems would undoubtedly have still been developed.

Seen from a communication perspective, it may strike some as odd to include television in this overview of media constellations, as today's general experience of the medium is predominantly dictated by unidirectional broadcasting of entertainment, news reports, and the like. However, this conception of television as 'radio with pictures' is but one of several possible dispositifs that were contemplated during its various development phases. Robida's stories and illustrations show that the old desire to see at a distance did not only express itself in entertainment or surveillance scenarios, but was also intimately connected to future visions of interpersonal communication. Especially in the years before the 1920s, when the idea of broadcasting had not yet been fully institutionalised, perfecting point-to-point communication was thought to be the logical route for media development, and therefore enhancing the telephone with live and moving images was seen as a useful addition to this medium's ability to construct a feeling of being present (Elsner, Müller & Spangenberg 1994: 112, 122). Once the first working televisual devices were finally announced in the mid-1920s, the utopian fantasies of being able to see as well as talk to someone far away were given a renewed stimulus, both in popular imagination and in laboratory research. Some of the well-known movies of the 1920s and 1930s, such as UP THE LADDER (Sloman 1925), METROPOLIS (Lang 1927) and MODERN TIMES (Chaplin 1936), contain paradigmatic examples of how two-way televisual communication systems were expected to develop and become commonplace in many areas (Koszarski 1998: 129-131).[42] These fictional visions quickly engendered real-life counterparts; both AT&T in the United States and the Reichspost in Germany introduced 'Bildfernsprecher' (or 'picture phones') from 1936 to 1940 (Reuter 1990: 210). After a few years, however, these experiments were ended because of a general lack of interest from the public (Elsner, Müller & Spangenberg 1994: 126).

The troubled track record of the picture phone dispositif provides an exceptionally good case in point that myths of improved communication can linger on in the course of media evolution, but do not automatically transform day-to-day telecommunication practices once the necessary components for realising those myths can be manufactured technically.[43] Although up to today there have been countless other attempts in many countries at launching vision-enhanced telephony, and although future visions of communication continue to stress the *topos*, videophoning continues to be used in niche markets only.[44] Still, myths persist, in dead, near-dead, and in living media. Instead of in the addition of moving images to telephony, television found its strongest supervening social necessities in the same domain as where radio had been operative since the end of the 1920s, namely the broadcasting of news bulletins, entertainment, sports events, and educational programmes. As with radio, a 'fantasy [...] of imaginary unity with "absent" others' sprang from the use of television, based on the idea that millions of people watched the same programme or were aware of the same events happening in the world (Spigel 1992: 116). The feeling of 'we-ness', which was already being broadcast across countries and continents by radio, was now visually beamed live into the living room on several channels. According to media theorist Rudolf Arnheim, one of the early scholars to theorise the medium of television, this visual extension of our social network could bring about the awareness of 'the place where we are located as one of many', making us 'more modest, less egocentric' (Arnheim 1957: 194). Arnheim was not the only one to voice such expectations; in his book *The outlook for television* (1932), Orrin E. Dunlap refers to television as a 'new medium of friendship', one that would obliterate suspicions and let 'the inhabitants of the earth [...] see how their fellow men live on the other side of the globe' (Dunlap 1932: 229). Here is the familiar adage that to get to know the other is a good thing, and that technology provides the binding glue; it is the belief that the harnessing of electricity, the conquest of the æther and the use of the broadcasting properties of radio and television would eventually lead to a society free from prejudice and misunderstanding, dissolving all barriers and thus advancing utopian images of social harmony, democracy, and world peace.

It should not be a surprise to see that all these hopeful expectations eventually subsided. Especially in the United States, where the leading radio networks (which were predominantly in control of television's development) did not want to repeat the same mistake they had made in initially failing to recognise radio's commercial potential, television had been destined to broadcast entertainment, not to function as a unifying communications tool (Boddy 1990: 16). Entertainment is now one of the core businesses of television networks in many countries all over the world and, despite globalisation trends, television often reproduces and stresses national identities that are taken for granted (Desaulniers 1986). The fantasies of being able to control whatever images were displayed on the screen also turned out to be illusory; not only did television prove to be a much more difficult device to build than the radio, restricting the scope of conceptual contributions by amateurs, but commercial networks and state-related institutions invariably took full governance of broadcasts, limiting free access to production and distribution. Since the second half of the twentieth century, the television viewer has been engulfed by a myriad of channels, which has reduced the chances of sharing viewing experiences and maintaining a feeling of 'we-ness'. Television offers a global collection of views on local cultures instead of constructing a McLuhanesque global culture,[45] and today, with the growing popularity of digital television and personal video recorders, the individual experience of television is gaining even more momentum. Soon, though, television's myths about the unification of humankind returned in another new communication technology: the Internet.

The Internet constellation: all together now

With the advent of television in the 1920s and 1930s, the ongoing stream of innovations in telecommunications seemed to have accomplished, for the greater part, the age-old dream of being able to hear, speak, and see from great distances. Accelerated by the creation of the electric telegraph, the evolutionary process of media development – which over the decades had continued to engender new knowledge, new technical components, and new receptive environments in which further experimentation could take place – was now poised to extend the three main communicative senses (hearing, speech, and sight) to a global scale. Because audio and visual technologies transmitted live signals, and thus facilitated instantaneous and simultaneous access to people and information, they contributed greatly, in all their various dispositifs, to the momentum of myths of unison through improved communication.

As part of this ongoing process, as we saw in the utopian expectations of especially radio and television, the positive outlook on the proximity of social unification on a global scale was upheld, motivated by the positivist premise that the more one knows of one's place and language amongst many, the more humble and cooperative one becomes; the 'dream of re-establishing the pre-Babel "great human family"', as Armand Mattelart writes, 'is present throughout the history of the imaginary of communication networks' (Mattelart 2000: 20). In the second quarter of the twentieth century we can therefore point to a heightened expectation that the new electronic media could be used to enhance access to knowledge resources, and improve the exchange of ideas, opinions, signs, and symbols. Among the thinkers that advocated such positivist ideas was the Belgian Paul Otlet, often seen as 'a pioneer both of international organisation and of documentation' who laid the foundations of what we now would call information science (Rayward 1975: 3). An idealist but by no means an otherworldly dreamer,[46] in 1934 Otlet published his *Traité de documentation [Treatise on documentation]*, a voluminous work in which he proposed to create a documentation system that would be '1° universal in its scope; 2° reliable and correct; 3° complete; 4° quick; 5° updated; 6° easy to obtain; 7° compiled in advance and ready to be communicated; 8° accessible to a great number of people' (Otlet 1934: 6).[47] Otlet envisioned that interaction with the system would make use of a form of networked teleconferencing, in which new media such as film, the gramophone, radio, and television would provide communication and information to a far better degree than a book could offer. As he noted, these instruments would make us equal in perfectness and fullness as to 'God himself' (ibid.: 431).

Although Otlet's detailed technical description of his system would, even by present-day standards of technological development, make it an unfeasible project,[48] it shows he was very much swayed by the potential of the new 'ubiquitous, universal and eternal' media of his time to help advance humanity towards a 'divine state' of 'being everywhere, seeing everything, hearing everything and knowing everything' (ibid.). He was not alone in harbouring this almost religious, Teilhardian drive. The idea that proper education and information dissemination, managed by the newest telecommunication systems, automatically engender a better world was thriving, influencing many thinkers of the time. A similar technology-inspired belief in the need for a permanent world encyclopaedia as Otlet's, for instance, can be found with H.G. Wells, the English utopian social reformer who in 1938 wrote that the creation of a universally accessible and 'complete planetary memory for all mankind' would constitute 'a real intellectual unification of our race' and ultimately 'a way to world peace' (Wells 1938: 86, 88). Around the same time, and unaware of his European contemporaries, American engineer

Vannevar Bush valued the importance of improving information management in much the same way. In an article written in 1939, published in 1945, he proposed to build a personal networked microfilm system with which scientists could concurrently store and retrieve documents, and construct 'associative trails' between those documents (Bush 1945).[49] As he saw it, such a system might even evolve into an all-electric version and directly link to the brain, elevating 'man's spirit' by augmenting his 'limited memory' (ibid.).

We thus see that the assumed unifying power of telecommunications technologies, which had already been touted at the inception of the telegraph and had accompanied many subsequent technological paradigm shifts, remained to shadow and seep through into the consciousness of a diverse field of scholars, writers, scientists, and engineers in the twentieth century. The development of the transistor and of electronic digital computer technology after the Second World War decidedly increased the awareness that linking people and information through new media would be a good, even a necessary thing. This became adamantly clear in the 1960s, when time-sharing systems made it possible for several users to remotely work on the same computer, thus creating a configuration of information management very similar to the one proposed by Otlet. In that period, American computer scientist Joseph Licklider, the director of computer research at the Advanced Research Projects Agency (ARPA) who cited Bush as a great influence on his thinking, wrote a number of illustrious papers in which he proclaimed that computers would become personal devices (Licklider 1960), which would be connected to a galactic resource network (Licklider 1965), and would facilitate the meeting of many interacting minds (Licklider & Taylor 1968).

Licklider, together with his successor at ARPA Robert Taylor, was the fundamental initiating force behind the creation in 1969 of ARPANET, the interconnected science community computer network that was predominantly funded by the American military-industrial complex (Flichy 2007: 41-45). Concurrently, three other computer networks had been constructed (a nameless network at the National Physical Laboratory in England, ALOHANET at the University of Hawaii, and CYCLADES in France), and with a growing number of computers being connected, soon the idea of a network of networks was born. With the development and large-scale tests of a set of standardised language protocols (known as TCP/IP, designed by Vinton Cerf and Robert Kahn), this super network, or Internet as it later would be called, became fully operational in 1983. International communication, largely in the form of e-mail, thrived, as did the storage, retrieval, and exchange of information. So, the visions of Otlet, Wells, Bush, and Licklider were for a great part realised, but in a very particular way: because the designers of the system were also the users, there was no need to cater to the wishes of an external market; moreover, because there had been no governing body controlling the direction of development, the technological and communication imaginaries could simply run free, creating the scientific community's ideal (ibid.: 35, 63ff).

After access to the Internet had become commercialised in the middle of the 1980s, the number of host computers and users grew dramatically.[50] This growth could be observed first among the academic communities, and from the mid-1990s onwards among the general public, with the advent of personal home computers, Internet service providers, and the World Wide Web. The Internet is not the first or only technological platform with which people can cooperate, share electronic information or forge ad-hoc virtual communities,[51] but ever since the start of its exponential growth it has dominated the media agenda, becoming a synonym for the single most effective means to make enlightened progress towards a unified and better world (Mosco 2004: 91). The fact that anyone with a connection could freely participate in the social sphere of cyberspace, for instance, led some early observers to proclaim that the Internet could 'perhaps revitalize citizen-based democracy' (Rheingold 1993: 14), or even create a world in which 'all the sentiments and expressions of humanity, from the debasing to the angelic, are parts of a seamless whole, the global conversation of bits' (Barlow 1996). The physical world was to be enhanced (or, for futurists like Ray Kurzweil [1999], even replaced) by the virtual world, blending atoms and bits into a ubiquitous and information-rich mixture, accessible and manageable on the fly, by anyone, anywhere.[52]

In the second half of the first decade of the twenty-first century, the unifying aspect of the Internet became undoubtedly best and most visibly captured under the heading of 'Web 2.0', a hugely popular metaphor which effectively put forth the notion that the Web had been improved to a new version. The prevalent conception was that there somehow was an 'old' version of the Web that had its bugs and flaws,[53] and that those bugs and flaws had now been taken care of. The power of the Web 2.0 metaphor is such that it tends to blend the many perceptions of what kinds of improvements have been made into a single idea: despite the fact that the term Web 2.0 is also used to designate a technological shift in how Web pages are now increasingly being built in a dynamic way using pre-stored data (Boomen 2007), what has come to the fore as its dominant meaning is the idea that the Web has now become truly social, and that its users are now participating and collaborating in creative and knowledge-building activities on an unprecedented scale.[54] Notwithstanding the critique that the amassing of personal information for the benefit of creating an all-encompassing web of knowledge can lead to serious privacy intrusions,[55] terms like 'collective intelligence' and 'global brain' keep resurfacing in present-day Internet myths, and continue to emphasise that out of intensified cooperation and participation a more democratic and unified world will emerge (see O'Reilly 2005; Macy & Thompson 2011: 16).

Cooperate, unify, participate, blend. We can see telecommunication's favourite topoi at work. The Internet's connotations of connecting and disclosing, rooted in idealised ideas of communication that over centuries have co-evolved with media technologies, cause well-known glorifications of new media to be reproduced. They tout the coming of the 'Omega Point of perfection', as Mosco aptly remarks (2004: 75). This allusion to Teilhard de Chardin's vision of an all-encompassing knowledge singularity is no surprise, considering that some of the most persistent claims found in discourses of the Internet are that its liberating powers will fuse all available intelligence (Lévy 1997; Shirky 2010), give everybody equal opportunities to participate (Shirky 2009), and bring about the end of distance (Cairncross 1997; Friedman 2005). We have seen very similar proclamations when telegraphic, telephonic, radiographic, and televisual technologies were envisioned, developed, advertised, and used. The important thing to note here is not that these claims are far-fetched, but that they are part of a long tradition of regarding new communication technologies as the latest step towards the utopian unification of minds, of combining engineering, regulatory, corporate, and consumer voices in a strategic deployment of utopian discourse. To be sure, this does not mean that it has always been business as usual, that nothing really changes; on the contrary, especially the Internet and its capacity to remediate other media through the digitisation of media content shows that media have no stable identities, and that their dispositifs are continuously changed to suit new telecommunication paradigms (see Wu 2010). What does not seem to change much, though, is that new media are initially met with high hopes, before they become mundane and create room for the expectation of the 'next big thing'. The following chapter will delve into the details of the next big thing of today: mobile communication technology.

4. Mobile communication dreams

It just makes sense. The more ways you have to connect with the people who matter to you, the easier it is to stay close. (T-Mobile 2007)

Today, gathering from industry and user accounts, the utopian desire for improved communication is fulfilled to the fullest yet by mobile communication devices. Great expectations for what they can do to make our lives better than ever before – palpably visible in advertisements and press releases but also disguised in how users motivate their wireless communication behaviour – invariably rely on the familiar adage that new technologies will finally solve old problems. The devices have become extremely widespread in a very short time, and are represented as the seemingly logical, natural, and inevitable outcomes of the ideology of improved communication: unlike any other medium before, they let their users act as both senders and receivers of information, wherever they are, immersing them in a vast network of interconnectedness. While we should remain sceptical of such rhetoric, mobile communication devices, even accounting for all the diversity and complexity in their use, do have, by their sheer presence in numbers, a generalising effect on how communication is to be understood at the beginning of the twenty-first century: their pervasiveness has rendered them ordinary, and, in a sense, invisible; they become our environment as they seamlessly blend into almost every social activity imaginable, extending, as Castells et al. (2007: 126) write, 'the beat of life into ubiquitous interactivity'.

The present chapter zooms in on this most recent branch of communication technologies in order to apprehend how and where idealised ideas of improved communication find fertile ground in today's mobile communication's discourses. It will do so by building upon the main themes covered in the previous chapters – ideas of progress and utopia that are projected upon idealised conceptions of communication, which in turn express themselves materially in the evolution of media technologies – and by formulating an answer to the question to what extent the many manifestations of mobile communication technologies and their popular apprehension, reception, and societal justification employ discourses that harbour forceful fantasies of a progressive path towards ultimate communication.

First, I will pick up the project of uncovering myths of improved communication in the evolution of communications technologies, and continue to trace those myths within the various developmental strands of mobile wireless communication up to the present day. As with the accounts of media developments described in chapter 3, the focus will not so much be on giving a dry enumeration of technological innovations, but more on how the environmental factors, which facilitated the evolutionary developments of mobile wireless communication, were shadowed and oriented by myths of ideal communication. This will be a short history of the framework within which beckoning and onrushing futures set the agendas for creating mobile communication devices. As we will see here, myths serve to strategically create the impression that there is little difference between what developers of mobile communication technologies consider to be the right (and therefore profitable) road to travel, and what the general public are offered as the latest and 'naturally' improved technologies to accommodate their communication needs.

Next, the discussion will shift to analysing the four characteristics that are typically ascribed to wireless communication technologies by advertisements and user accounts, in order to gauge the extent to which idealised ideas of improved communication are articulated in present-day discourses of mobile communication. These four characteristics are comprised of the ability to facilitate ubiquitous connectivity, fluid sociability, real-time relief of anxiety, and omniscience and collectiveness through ever-present knowledge. I will explicate how both the fulfilment and the unintended consequences of the longing for a communication sublime can be located in the various deployments and appropriations of these characteristics, and subsequently pinpoint the paradoxes that become articulated in the incessant drive for communication improvement. As will become clear, because of their ruthless and pervasive connectivity, mobile communication devices make us painfully aware, perhaps more so than other communication media, that the actual achievement of ideal communication is both a blessing and a curse.

The final section of this chapter will show that the mobile communication condition that we live in today is itself, of course, not exempt from the notion that there are still communication problems to be solved. By discussing two of the recent trends that are positioned as representing the next important phase in the ongoing search for improvement in the development of mobile communication technology, namely location awareness and the creation of an 'Internet of Things', I argue that myths of ideal communication continue to orient our thinking about new communication technologies, and are invariably translated from older forms of myths so to have them work as solutions for present-day communication problems. Mobile communication technologies are here to stay, but they are definitely not the final answer.

Making communication mobile

> **Dick**: *I'll be at 362-9296 for a while; then I'll be at 648-0024 for about fifteen minutes; then I'll be at 752-0420; and then I'll be home, at 621-4598. Yeah, right George, bye-bye.*
>
> **Linda**: *There's a phone booth on the corner. You want me to run downstairs and get the number? You'll be passing it.*
>
> (PLAY IT AGAIN, SAM, *Woody Allen 1972*)

The development of mobile communication technologies is one that cannot be isolated from historical, social, economic, and political contexts, nor from the dreams of perfected communication that pervaded those contexts. Moreover, the lesson learned so far is that the evolution of electronically mediated communication knows a myriad of interconnected branches, some still extending themselves and others long buried under the weight of history. Within these branches we can locate numerous human and non-human actors, which at some points bred the most fantastic but stillborn offspring and at other points merged to give life to contraptions that hit the sweet spot of contemporary communication desires. The roughly 150 years that span between the commercial deployment of the electrical telegraph and that of the Internet saw many such failures and successes, and the previous chapter showed that all of these relate in various ways to notions of wanting to remove obstacles that stand in the way of perfectly communicating from afar, of making us omnipresent and omniscient. Within this vast forest of old, new, living, and dead media, we should therefore also be able to find the areas that harbour the earliest technological roots of personal wireless communication.

Envisioning personal radio communication

While fantasies of direct, telepathic-like communication are certainly not confined to the last few centuries, it was not until the mastering of electricity and the development of electronically operated communication technologies that the first feasible ideas were put forward regarding how actual wireless person-to-person communication might work. Within the lengthy history of experimenting with the electromagnetic spectrum and designing devices that could exploit its characteristics, the start of the twentieth century marks a time when two powerful media constellations had started to approach each other. The first of these two, comprised of telephonic dispositifs, had already begun to present itself as a fierce competitor of the telegraphic constellation. The telephone noticeably made its way onto the evolving telecommunications scene in the 1880s and 1890s, and had gained a healthy momentum. Its ease of use, coupled to a simple but smart rental scheme pioneered by the Bell Telephone Company and continued by AT&T under the helm of Theodore Vail, quickly captured the attention of those who needed swift communication and could afford it (Aronson 1977: 27). At first, subscribers were predominantly rich people who worked in professional and commercial industries, but with the expiration of Bell's key patents in 1893 and 1894, many small telephone start-ups in the United States began to offer telephonic services at greatly reduced rates, making them available to the general public as well (Fischer 1992: 42-44). In 1914, around ten per cent of the United States population had a telephone; in other developed countries, the figures ranged from 6.5 per cent in Canada to 1.7 per cent in Great Britain (Kingsbury 1915: 530-531). Although far from being the near-global medium it is today, for many at that time the telephone was a well-known device, symbolising the next significant step in the direction of the presence of an uninhibited 'freedom of expression' of mind (Boettinger 1977: 203).

The second media constellation that signalled a change in late-nineteenth century's telecommunication discourses presented itself in the form of radiographic experiments and devices. Still in a phase where the understanding of the wireless transmission of electromagnetic signals was rudimentary, these new imaginative conceptions of bridging space by technological means were poised to fully interact with the dominant telephonic dispositifs that had been established. As evident from the William Crookes' article discussed earlier, in 1892, wireless point-to-point communication seemed to be on the brink of invention, and the prediction of its onset was to be repeated many times, in spite of the fact that later on wireless signalling revealed itself to be much better suited for broadcasting purposes. In a wonderful collection of forecasts brought together by Thomas H. White, we find, for instance, that in 1901 a commentator in the *London Spectator* saw that '[s]ome day men and women will carry wireless telephones as today we carry a card case or camera', and that in 1902 the English journal of electromagnetics *The Electrician* reported

> that 'a number of scientists scattered all over the civilised world are eagerly seeking the solution to the problem of wireless telephony', and although so far there had been only limited success, 'A future generation may conceivably accomplish as much in wireless telephony as is dreamed of to-day by visionaries'. (White 2007)

The persistence of this belief was closely linked to the idea that all that was needed was new technologies; while the concrete practices of wireless signalling at the beginning of the twentieth century were far removed from the dreams of visionaries, further development of radio components like the crystal detector and vacuum-tube transmitters continued to improve radio's potential and therefore fuel the public and scientific imagination. As Pool et al. revealed from a 1910 article by Herbert Casson, one particular inventive notion was that police work would benefit greatly if 'each individual [would] have a number by which he could be reached telephonically wherever he might be' (Pool et al. 1977: 138). The wording closely resonates with the familiar desire for omnipresence and as a result omniscience through improved communication, something that seemed to be virtually present merely by cleverly combining available technology. Over the years, little would change in this expression of this desire; White writes for example that in 1919, in the U.S. War Department annual report,

> Signal Corps head Major General George O. Squier talked of 'the day which I believe is not far distant, when we can reach the *ultimate goal* so that any individual anywhere on earth will be able to communicate directly by the spoken word to any other individual wherever he may be'. (White 2007, emphasis added)

The same words, the same wish; out of the ideational entanglement of the telephonic and radiographic constellations came many similar predictions that echoed hopes for ever-more ubiquitous communication. As is so often the case, however, the day that the ultimate goal was to be reached would always be further away than projected. While many dreamed of portable radiophones at the end of the nineteenth century, the reality was that, for a long time, communication on the move was only possible if one had a large enough sea or land vehicle to carry the heavy radio equipment.

Vehicular mobility

The first successful commercial deployment of wireless communication was, in fact, by its very nature mobile, as Marconi had attracted the British, Italian, and American navies as well as the maritime insurance company Lloyd's as main customers of his wireless telegraph systems. He competed with wired telegraph networks on land as well, but once he had begun to adapt his technology to the maritime environment in 1897, there was an increased demand for ships to be fitted with radio equipment (Dunlap 1937: 76-83). After the advantages of having a reliable means to communicate over long distances had become evident in naval conditions, Marconi sought to expand the potential market for mobile communication to other areas. In 1901, the same year he managed to send a radio signal across the Atlantic for the first time, he equipped a steam-powered wagon with a transmitter, receiver, and electric batteries so that 'communication [could] be maintained while the vehicle [was] travelling' (*Western Electrician*, July 27, 1901: 51 cited in White 2005). There was no immediate and direct financial gain made from this experiment, but an important connection had been established: this was the world's first car-based communications system, and as such it is a typical example of how the branches in the evolution of communication media, guided by experimental or entrepreneurial motivations, continuously find new areas to grow towards.

Around the same time, and emphasising the argument that while the technological means to communicate evolve perpetually they have a tendency to converge towards similar 'good tricks', we see another car-bound experiment in mobile communication, albeit one without radiographic technologies: in 1910, Lars Magnus Ericsson (the Swedish manufacturer of telegraphs and telephones) and his wife Hilda built a telephone system into their car, and by physically attaching that system to the wires of telephone lines using two long grappling poles, they were able to stop along the road and make a call wherever they could make a connection (Agar 2003: 8-9). Despite the similarities, there was a significant difference between this car-bound system and that of Marconi, however, because even though the technique of 'hooking up' a telephone to existent wires provided ample communication advantages – it had also already been successfully employed by armies in battle and by telephone companies to test their lines – it only provided mobility *between* fixed communication; it was the radio-based system of Marconi that made mediated communication *itself* truly mobile. Within the general perception of what mobile communication was or could become, for a long time technical affordances as well as prevailing social necessities would retain that link between automobiles and radio equipment.

The supervening social necessities for cars that offered mobile communication capabilities first appeared in the form of a desire to improve law enforcement. One optimistic forecast that had been made during the fixed telephone network's development was that crimes would be solved more easily, because many more people would be able to quickly report them to the police. This brought forward the need to have police officers on call as much as possible while on the beat. With the technology at hand to instantly contact anyone with access to a radio receiver tuned at a specific frequency, the practical solution to the problem of the need for increased accessibility of the police became straightforward. From 1921 to 1928, several police officers, all radio amateurs, developed a mobile radio communications system for the Detroit Michigan Police Department, which they fitted in their cars (IEEE 2007). Initially, this system, which in the early phases relied on the use of Morse code but later provided fully voice-based communication, was one-way only, so that alerted police officers had to stop their cars and use a fixed line to call in (Gow & Smith 2006: 23). It proved to be very successful nonetheless, and more police departments quickly followed suit. This technological momentum was given an extra impulse after the Galvin Manufacturing Corporation (GMC) produced a two-way radio system under the brand name 'Motorola',[1] which became the land-mobile radio of choice for many government agencies and emergency services during the 1930s (Goggin 2006: 25). The subsequent deployment in the Second World War of the GMC-manufactured AM portable two-way 'Handie-Talkie' in 1940 and the FM portable two-way 'Walkie-Talkie' in 1943 showed the immense and often strategic usefulness of being able to communicate back and forth without the restrictive need for a fixed connection, and paved the way for larger-scale commercial experiments with car-based radio telephony after the war.

These post-war experiments were not without their technical difficulties, which were of course discursively framed in terms of challenges that had to be overcome: the problem of requiring wires may have been put out of the way, but there were other obstacles to uninhibited communication that loomed at the horizon. Up to the mid-1940s, radio communication had in effect relied on a rather closed system: you could only talk to someone who used proper radio equipment and was tuned in on the right frequency. Wireless calls to a fixed phone (or vice versa) could not be made,[2] and the frequency range in which radio communication worked was limited. Although AT&T and GMC both tried to address all of these limitations, resulting in the creation of the Mobile Telephone Service in 1946, radio channels in the available radio spectrum were sparse and suffered heavily from interference; in addition, the service was costly, the equipment was heavy, and conversations could only take place using a push-to-talk procedure (Gow & Smith 2006: 23-25).

What made things still more difficult was that, despite all these drawbacks, the demand for car telephones in the two decades following the Second World War was overwhelming; many people who wanted to experience mobile communication were placed on a long waiting list (Agar 2003: 36-37). Especially for those who were often on the road, the potential communicative affordances of mobile telephony far outweighed its technical imperfections. This pressing urge to make use of mobilised communication manifested itself in many countries, and for a variety of reasons. In 1949 in the Netherlands, the government-operated Postal, Telegraph, and Telephone agency created a national public mobile network that used centralised base stations to establish contact between any telephone and 'mobilofoons' (Meulstee 2007). This network was predominantly used by taxi services and transport companies, and incidentally proved to be of great importance for the coordination of emergency services during the 1953 North Sea flood (Museum voor Communicatie 2006). In Sweden, Ericsson built the world's first fully automatic Mobile Telephone A system in 1956, which attracted the patronage of many lawyers, doctors, and people living and working in rural areas (Agar 2003: 49; Ericsson 2006). Thus, although it was used mostly in specialised fields and its costs were much too high for the majority of people to own one, the personal mobile communication device exerted a powerful attraction on the collective imagination, and had a large appeal for many who saw connectedness as essential for their everyday functioning.

As had been the case with the fixed telephone, the growing popularity of mobile communication increasingly put pressure on its infrastructure, creating a dire need for further innovation. It was Bell Laboratories, which since 1925 had represented the institutionalised search for improvements in communication, that was instrumental in providing the necessary developments and laying the foundations for the ubiquitous existence of present-day mobile telephones (Goggin 2006: 26). One of those important developments was the invention of the transistor in 1948. It made the miniaturisation of radio devices possible, and as such transformed radio telephones into much less bulky devices. The second innovation was the idea for a cellular network, in which multiple and interconnected low-powered base stations organised in a decentralised hexagonal grid would replace the few centralised high-powered radio towers, thus freeing up frequencies. Conceived in 1947, it would take more than twenty years before the cellular system would become operational because of two reasons: first, the American Federal Communications Commission (FCC) refused for a long time to allocate the requested frequency spectrum, and second, the technology that was needed to manage call handovers and frequency switching was only developed in the late 1960s (Farley 2006b, 2006c). Still, the cellular system proved to be key to the mobile system's success.

Leaving the car behind

In the late 1960s, while AT&T focused on using cellular technology to improve car telephony, Motorola (which had changed its name from Galvin Manufacturing Corporation in 1947) aimed to extend the mobile telephony paradigm beyond the car. To Martin Cooper, the general manager of Motorola's Communications Systems Division at the time, the advantages of a truly portable telephone were unambiguously clear: 'When you park your car and leave, you can't use your mobile [car phone] but you can take your portable with you' (Motorola 2007). Cooper thus described what he called the 'original dream' of mobile wireless communication in well-known terms: to be able to use a communication device wherever and whenever, and not have it be bound by any physical object other than the human body itself (Charny 2003). Notably, Cooper reframed the 'reverse salients' of his project, the very concrete and historically contextualised technical problems that were his daily concern, as standing between him and realising an apparently long-held dream of unlimited communication. His determination to realise this dream vividly reflects how myths of improved communication resonate through Western culture, and how they find fertile ground with many people, including – especially – those who have a say in the development of new communication technologies.[3] Cooper took free rein and, within five years after the FCC's 1968 declaration that if cellular technology proved technically feasible it would be allocated a large number of frequencies, he and his team had designed and built a prototype of the DynaTAC, a brick-like portable phone weighing almost one kilogram. It would take another ten years and many FCC hearings before Motorola's first commercial portable cell phone, the DynaTAC 8000X, was approved and made available in 1983.

Cooper was not the only one who had the dream of creating boundless mobile communication; the topos flitted through many a telecommunication designer's brain, and this can be seen in the rapid advances made elsewhere in the world. While the cellular concept and the necessary technologies to make cell phones work were developed in the United States during the late 1960s and early 1970s, their commercial deployment was hampered by legislative restrictions and high costs to such a degree that other countries implemented successful cellular networks a few years before the Americans did. In Japan, the Nippon Telephone and Telegraph company established a cellular service in 1979 (Goggin 2006: 29). Then there was the Nordisk Mobil Telefoni system, which was put into use in the early 1980s in Sweden, Denmark, Norway, and Finland (but which, curiously enough, premiered in Saudi Arabia) (Ericsson 2007). The rest of Europe followed closely after the Advanced Mobile Phone System had been inaugurated in the United States in 1983, with the United Kingdom installing its Total Access Mobile System, Italy its Radio Telephono Mobile Integrato, France its RadioCom 2000, Germany its C-Netz, and the Netherlands their second generation Auto Telefoon (ATF-2)[4] network (Gow & Smith 2006: 46; Groen 2006; Ling 2004: 8-9). By the end of the 1980s, mobile telephony had gained a momentum comparable to that of fixed telephony in the 1910s: the infrastructure had undergone a major change in order to service more subscribers (like automated exchanges had done for wired telephone networks decades earlier), and although not everybody had the financial means to actually own a mobile telephone, the portable technology had firmly established itself in the public awareness as the next generation in the evolution of telecommunications.

Global expansion

Still, the ideal of universal mobile telephone service that Cooper and others had dreamed of was far from being fulfilled. It was the increasing amount of waiting potential customers that made this abundantly clear, emphasising that the idea of personal wireless communication had a broad level of appeal. Despite the considerable costs, the demand for mobile telephony turned out to be much greater than expected, and soon after the first generation of cellular networks had started operating, 'predictions of unlimited capacity [...] proved grossly optimistic' (Fox 1990: 45). Another factor that hampered universal service was that, because all the different cellular systems were incompatible with each other, it was practically impossible to use a phone abroad; only countries that shared a similar system could provide roaming access. Especially in Europe, where both Enlightened unification ideals and a drive towards global economic competitiveness determined much of the political and industrial agendas of the Union's member countries, this incompatibility was a thorn in the side of many; therefore, the project to 'build one European cellular phone system [...] would be a major material means of realising the dream [of unification]' (Agar 2003: 58). The subsequent development and implementation of the Global System for Mobile Communications (GSM) in 1991 was completely in line with the European commitment to this dream: it was adopted as a pan-European standard, lifting cellular systems from the first (analogue) to the second (digital) generation and freeing up spectrum in order to manage the rapidly increasing amount of subscribers (Ling 2004: 9). Concurrently, in the Americas and parts of Asia, newly developed second-generation digital cellular telephone standards started to employ Code Division Multiple Access protocols, and in Japan the switch was made to the Personal Digital Cellular standard (Gow & Smith 2006: 48-49). All in all, at the beginning of the 1990s in many countries around the world, important technological improvements had been made to accommodate for a lot more mobile telephone users than the first generation of cellular systems could have managed.

However self-evident this process of mobilisation might seem, it should be carefully noted that, even though a lingering desire for media that could establish personal wireless contact had existed for a long time, it was only when particular contributing factors and actors[5] came together in the media environment of the early 1990s that the mobile communication condition as we now know it began to take concrete shape. When it did, we see that, from the mid-1990s onwards, a rapid increase in the amount of mobile telephone subscriptions took place.[6] Although the transition from an analogue to a digital system was instrumental for this to happen, it was not the only reason why mobile telephones grew in popularity. With further technological developments and components becoming smaller and cheaper, mobile telephones had transformed from clumsy heavy bricks to elegant and even fashionable portable objects, making owning and using one much less of a hassle – or an embarrassment. No company was more prominent in the early days of the production and marketing of such small 'personal trusted devices' than Finland-based Nokia, and it had understood very well that the desire to be constantly in touch was not just to be found among people doing business. Recognising the demands of regular consumers was the first step and, accordingly, designing mobile telephones the next. Nokia's ensuing segmentation of the mass market into a broad range of lifestyle-related niche markets proved to be enormously successful, and paved the way for the unprecedented adoption of mobile telephones amongst a huge range of customers (Steinbock 2005: 167).

Today, mobile communication devices are virtually ubiquitous.[7] Their ability to potentially connect anywhere, anytime, to anyone or anything in the informational network, combined with an ongoing emphasis of the individual as the nexus of communication and entertainment, has sculpted the devices into vital cultural artefacts, providing many more functions than just placing calls. A significant proportion of devices nowadays supports text messaging, playing games, taking pictures or video clips, and listening to radio or digitally stored music; the more advanced models enable activities such as surfing the Internet, reading emails or other text documents, managing all kinds of social media hubs, using Global Positioning System (GPS) applications for location-based services, and even paying money. Any social activity may now involve using one or more of these mobile functions at some point, suffusing our understanding of communication with notions of immediacy, ubiquity, intimacy, reassurance, and knowledge gathering, all potentially available at the press of a few (virtual) buttons.

In the evolution of media technologies, mobile communication devices constitute its more recent branches, and their 'magical fetish' (McGuigan 2005: 46) engenders dreams of improved communication very much like previous generations of communication technologies did. So much so, in fact, that for some scholars it is tempting to describe the rise of the mobile society as not only revolutionary, but also almost as a coming home to the 'true' balance of communication which is supposedly hardwired in humankind.[8] Mobile wireless communication technologies, in this view, constitute the ultimate means by which the longing for improved communication can be fulfilled today. However, such optimism, while partly justified by the sheer speed with which mobile communication devices have proliferated themselves and have helped many to improve their businesses or social lives, is not entirely warranted, as it underestimates the inevitable problematic consequences of the search for ideal communication. Just as with earlier attempts – or even more so because of their unprecedented ubiquity and pervasive nature – mobile communication devices will stress the dark side of all that an achievement of perfected communication entails.

Living the paradoxical dream[9]

Mystery Man: *We've met before, haven't we.*
Fred Madison: *I don't think so. Where was it you think we met?*
Mystery Man: *At your house. Don't you remember?*
Fred Madison: *No. No, I don't. Are you sure?*
Mystery Man: *Of course. As a matter of fact, I'm there right now.*
Fred Madison: *What do you mean? You're where right now?*
Mystery Man: *At your house.*
Fred Madison: *That's fucking crazy, man.*
[Mystery Man hands Fred a mobile telephone]
Mystery Man: *Call me. Dial your number. Go ahead.*
[Fred dials the number and the Mystery Man answers]

Mystery Man: *[over the phone] I told you I was here.*
Fred Madison: *[amused] How'd you do that?*
Mystery Man: *Ask me.*
[Fred remembers the anonymous video tapes]
Fred Madison: *[angrily into the phone] How did you get inside my house?*
Mystery Man: *[over the phone] You invited me. It is not my custom to go where I am not wanted.*
Fred Madison: *[into the phone] Who are you?*
[Both Mystery Men laugh mechanically]
Mystery Man: *[over the phone] Give me back my phone.*
[Fred gives the phone back]
Mystery Man: *It's been a pleasure talking to you.*
(LOST HIGHWAY, David Lynch 1997)

The important question that confronts us at this point is how exactly the discourses and infrastructures that co-construct current-day mobile communication dispositifs sustain idealised ideas of communication. How are these ideas expressed, to what extent do they influence everyday experiences of mobile communication? In one rather straightforward sense, the speed alone with which the adoption of mobile technology takes place suggests that there is a significant overlap between what people consider to be their natural communication needs and what mobile communication media have to offer. A quick glance at the breakneck-paced integration of mobile communication devices into everyday life could easily give the impression that the age-old desire to bridge distances and instantaneously communicate with anyone and anywhere, which has expressed itself countless times in predictions and glorifications of new media, is now for many people closer to fulfilment than ever before. Without truly realising it, we might already have arrived at the future of communication – as we have always seemed to be doing, as victims of our wilful amnesia.

But such an explanation would obscure the different reasons why people in various demographic categories adopt mobile communication technologies. These adoption strategies in their turn depend on a broad range of cultural, social, economic, and political settings, which also cannot be ignored as influencing factors in the construction and propagation of representations of idealised communication. So the question becomes this: while looking at analyses and discourses of everyday practices of mobile communication, do we indeed predominantly find notions that we have established a true communications utopia? Or is such a utopia effectively out of reach, despite the ubiquitous presence of the best candidates so far to satisfy the basic technological requirements for it to exist? Mobile communication devices may indeed facilitate the kinds of contact that perfectly fit idealised conceptions of the zenith of communication, but do we experience them as such? In other words, what are the kinks in the ideals, where is the uncanniness, what are the communication paradoxes in the mobile age?

To answer these questions, analytical cuts will be made along four characteristics of current-day mobile communication devices that are typically presented as their strongest selling points, which are identified in their promises to bring ubiquitous connectivity, fluid sociability, real-time relief of anxiety, and omniscience and collectiveness through ever-present knowledge. In each of these analytical cuts, the focus is first on how notions of improved communication manifest themselves, and then on the ways these various expressions and expectations of idealised communication run into the paradoxes of utopian and universalising thinking. In order to gauge user experiences from a micro-scale perspective, these analyses will engage with illustrative examples of research that empirically measured sociological and psychological ramifications of the use of mobile communication devices.

Ubiquitous connectivity

The obvious and single most defining characteristic of mobile wireless communication technology, one that precedes and co-defines its other specific features, is that it renders space largely irrelevant as a variable in constituting mediated contact. There is no need to be in a certain fixed location in order to connect to someone or something; given a well-diffused infrastructure that supports the sending, carrying, and receiving of radio signals, and provided that one is enveloped in one of the hertzian bubbles thereby created, a mobile call or a wireless Internet connection can be made anywhere. Increasingly, these bubbles are covering large parts of the Earth, from urban regions to deserted plains, mountain tops and out at sea. Due to contributing factors such as the standardisation of communication protocols, the ease of construction of basic technological frameworks, the portability of devices, the intuitive use of mobile telephones, and the high social and cultural value of personal communication devices, wireless services steadfastly have become globally pervasive indeed. What is significant about this process is that the supporting infrastructure is often rendered almost invisible, in a conscious attempt to create and uphold the illusion that the wireless connection is 'just there', to be invoked at will to magically synchronise different space and time coordinates.[10] Similar to the early seventeenth-century fantasies of magnetised compass needles that would move in communicative rapport wherever they were, a certain sense of – and need for – telepathic immediacy pervades modern wireless communication technologies: just turn on the mobile device, and a connection will be guaranteed to exist almost instantly.[11]

With their near omnipresence, mobile communication devices facilitate the further compression of geographical space into what Manuel Castells calls the 'space of flows', a concept he developed in *The rise of the network society* (1996), and which he defined in later work as 'the material organization of simultaneous social interaction at a distance by networking communication, with the technological support of telecommunications, interactive communication systems, and fast transportation technologies' (Castells et al. 2007: 171). In the space of flows, people, goods, and information are in a constant state of flux, moving between physical locations while being part of a dynamic network that is linked together through the use of communication technologies. It is connections, not places, that constitute the networks of the space of flows, redefining space 'into the space of communication' (ibid.: 178). What Castells et al. rightly observe is that this process of emphasising communication over location is accelerated even more today. Because mobile communication devices radically alter the long-lasting relationship between communication nodes and fixed locations, spatial vectors in the space of flows become increasingly heterogeneous, and consequently simultaneous social interaction at a distance turns into a pervasive activity that can be engaged in anywhere, at any time. The space of flows does not completely lose all sense of place (people continue to use communication technologies to make arrangements to meet each other, for instance), but places have become less significant in the establishment of communication now that the connections that are required to set up communication have been uncoupled from physical locations. Many may call the devices 'mobile' telephones, and ubiquitous connectivity does indeed support mobility, but it is the omnipresent availability of connection nodes that actually defines current-day wireless communication.

What follows is that, when distances and the need for location-specific contact points are removed as obstacles hindering what is perceived as the ultimate goal of unchecked communication, we become more and more immersed in what social psychologist Kenneth Gergen calls the 'relational net', in which everyone and everything can potentially link up (Gergen 2003: 111). Having a wireless communication device like a mobile telephone at one's disposal implies having access to ever-present, real-time communication channels, and thus to the means to engage in dialogue or to disseminate information whenever and wherever one wants. A mobile telephone therefore provides a very strong psychological 'fix' by supplying an abundance of communicative choice, the freedom to electronically connect and mediate knowledge, opinions, and desires; it is an apparatus of opportunity, a potential-rich portal. As such, it resonates with positively connotated notions similar to those found in the lure of the new and in the ideas of improving communication: as long as there is the chance that things can be turned around or improved, preferably by purposefully creating or using enabling technologies, there is the necessary illusion that we have a say in our destiny. Communicating has always been a fundamental social activity in human existence, and the idealised drive to improve the means to do so posits itself as an integral part of media evolution, so it should be no surprise to see that discourses of wireless communication technologies exhibit very familiar ideographs such as 'opportunity', 'hope', and 'progress'.

Some empirical examples might serve to clarify the above observations. The tendency to depict mobile telephones as powerful devices that possess an unlimited connective potential, and the habit of conflating this capacity for connectivity with the idea that it best answers our longings for the ability to create a better future, are, for instance, reflected by the strong emotional and cognitive investments that people make in wireless communication technologies. In many countries they are called 'hand phones', or their names refer to the fact that communication is 'always on' and can always be carried along (Thompson 2005: 17-19; Townsend 2002: 68-69). The most dramatically expressed urge to attribute magical powers to the devices is found in Israel, where a mobile telephone is called *pelephone* (after the name of Israel's first cellular communications company) which literally means 'wonder-phone' (Thompson 2005: 21). One can gather from these naming conventions that mobile telephones are considered essential extensions of the body. Moreover, mobile telephones and their invisible but potentially present connections have now become so intimately integrated into our being that many people experience feelings of panic when they find they have not brought their mobile telephone with them, or think it is lost (Vincent 2003: 220; Baron 2011). The severance of the hertzian umbilical cord is felt like an amputation that many prefer to avoid. To illustrate this further, during the short-lived offer in 2000 by several mobile telephone operators to provide free calls in the evenings and on weekends, some people even chose to maintain a constant connection through their mobile telephones, and listened to each other sleeping (Licoppe 2003: 177).

From a socio-psychological standpoint, mobile communication devices can be said to be experienced as the latest candidates to gratify the wish to come closer to communication utopia. They do so, however, not only because 'improved communication' has long been portrayed and perceived by producers and consumers as the one thing that can actually propel us forward into a utopian future, but also because the very idea of what constitutes 'improved communication' has co-evolved with media evolution. Chapter 3 showed that idealised ideas of communication have been expressed in a multitude of ways in a plethora of media dispositifs, but in the specific form of small, light, portable, and fashionable mobile communication devices, these ideals have now taken on the shape of the desire to always be in control of all available information flows, to be a perpetually connected node in an information-rich network. Mobile communication devices thus not only represent the most recent bid in a long line of recurring and similar attempts to improve communication, they also show how material and cultural aspects of technology co-determine dominant contemporary ideologies of communication media.

But what does this mean for the understanding of what 'communication' is? Has it indeed been improved? What are the unintended consequences of increased connectedness, what are the kinks in the ideals? At first, a perpetually connected state seems to be a perfect point of departure for establishing and managing all kinds of successful communication situations, and thus for reaching the ideal goal of a common understanding. Indeed, all of the scenarios for improving communication discussed in chapter 2 rely on the presence of channels – the essential prerequisites for any type of communication to exist – and, as the line of thinking goes, what would be a better improvement than making sure that those channels are available and operative at all times? But a closer inspection of what is at hand reveals that merely increasing the opportunities to connect does not suffice, and paradoxically might not even help in bringing people closer together in a utopian fashion. Through an analysis of several mobile telephone advertisements, press releases, and news reports, and by contrasting those discourses with ideas of communication found in works by Martin Heidegger, Jürgen Habermas, and George Myerson (2001), for instance, comes to the conclusion that what he calls the process of 'mobilisation' is predominantly geared towards just making contact, and not particularly towards creating a shared understanding. Although he is prone to overestimating the significance of press mumbo-jumbo by taking its seductive scenarios too literally as accurate representations of how people see and use their mobile telephones in everyday life, Myerson does pinpoint the industry's unidirectional sense of what constitutes ideal communication. He shows that, despite the fact that more and more people have the opportunity to engage in meaningful conversations, mobile telephones are marketed as personal communication centres with which one can satisfy individualistic wants, simply by making 'basic contacts' (Myerson 2001: 27). He argues that by the very ease with which mobile telephones can facilitate contact, the focus – in an almost Shannon-like sense – has shifted away from the content towards the channel of communication, a process that in his view degenerates our notion of what communication is. By reducing the idea of communication to the mere act of making contact, Myerson maintains, we actually lose touch with each other instead of sharing our sense of being (ibid.: 58).

And indeed, there are research findings that can be interpreted as supporting Myerson's claim that in the mobile age 'communication' is quickly supplanted by 'making contact'. Studies have shown that people generally take less time calling on mobile telephones than they do on fixed phones (Licoppe 2003: 175), think of asynchronous text messaging as 'quicker and more convenient than voice telephony' (Ling 2004: 150), and sometimes lock themselves in 'tele-cocoons' from which they only keep in touch with their most intimate friends, refraining from communicating with the outside world (Ito, Okabe & Matsuda 2005: 10-11). While some of these phenomena can be explained as user strategies aimed at managing billing systems and keeping costs low, the suggestion remains that mobile telephony and its infrastructure offer affordances that primarily invite what Christian Licoppe calls a '"connected" practice of interaction' instead of a '"conversational" practice' (Licoppe 2003: 174, 183). The 'anywhere, anytime' paradigm so much pervades interaction in everyday life, Licoppe found, that users prefer to think of their mobile communication practices mostly as intuitive whenever-you-feel-like events, which do not have to adhere to conventional communication habits and routines. As a result, mobile messages – especially text messages – are frequent but short, and often primarily serve to acknowledge the phatic dimension of communication, the connection that people share (ibid.: 180-181).

While such short communicative gestures are important elements in the shaping and maintenance of what can be called a 'performative value' in social relationships (Green 2003: 207), they do add to swelling flows of information, messages, and data, all of which affirm one's connectedness up to a highly redundant level. If always being connected is what brings pure communication a step closer, it is also what foregrounds the communication paradox, and forces us to realise that pure communication is relentless in its intrusive nature. Research has shown that when the times and places at which people can be contacted for whatever reason are extended to all possible environments, and those contacts become more frequent, a tension arises between the desire and expectancy for immediate and unlimited access to others on the one hand, and the need to filter out and restrict incoming access requests in increasingly variable circumstances on the other: the presence of a perpetual connection pressures users of mobile communication devices into managing all kinds of complicated communication schemes (Sherry & Salvador 2002: 114-115), into employing inventive ways to cope with awkward social situations caused by disruptive incoming calls (Ling 2004: 123-143), and even into being deprived of much needed sleep when they are kept awake at night by calls or text messages (Turrettini 2007). Judging from marketing predictions for next-generation wireless devices, such problems are not about to go away soon; the 'always on, real-time access' adage dominates the wireless industry's mentality towards the future of communication, reiterating along the way the wish of Licklider, Otlet, and countless others before them, for omnipresence and omniscience.[12]

Discourses of wireless communication technologies, then, reflect how the compression of space through ubiquitous connection nodes is of constitutive importance in the transformation of our understanding of communication. Across a wide range of demographic compositions, many social activities that used to rely on physical proximity or on the pre-arranged coordination of interaction are now reshaped into ad-hoc patterns of de-spatialised and heterogeneous contacts. Some have argued that the ongoing 'mediatisation' of the world blurs our understanding of social space and related activities to such an extent that we are left with 'no sense of place' (Meyrowitz 1985), or become entangled in the 'perpetual present' of a 'glocalised' world in which both global and local events and activities merely exist by virtue of their potential to be immediately interconnected (Virilio 1997: 129-145). The horrifying endpoint of this trip towards evermore connectedness, the argument goes, is one where everything collapses into a singular flow of being, much like the Borg Collective in the STAR TREK universe experiences everything all at once. In such a scenario, communication in its common sense ceases to exist; there will be no 'other' to relate to.[13]

The fact that such generalised outlooks on the negative effects of the loss of relational awareness can be countered by observations that mobile communication devices restore communal feelings previously eroded by mass media like radio and television (Gergen 2002), and that they do not by definition homogenise family members – let alone people from different cultures – in their communicative behaviour (Castells et al. 2007: 74-75), shows that mobile communication devices exhibit all the hallmarks of the dominant topos. In part oriented by idealised ideas of communication, new media appear to bring us closer to ultimate connectedness, but while doing so they trigger a wide range of ambiguous uses, appropriations, and behaviours. With the shift towards more connectedness, we thus experience faster and more often than before that the desire for pure communication brings us both pleasure and discomfort. As Michael Arnold asserts, the mobile telephone is Janus-faced; it is part of a socio-technical system that is 'not reducible to a direction or valence tipped with a single arrowhead, but better understood as a conflation of tangential implications, at least some of which can be read as ironically and paradoxically self-contradicting phenomena' (Arnold 2003: 234). Because of mobile telephony, the paradox hiding in the desire for ideal communication becomes more articulated than ever.

Fluid sociability

The observations on the strong significance of the connectivity aspect of mobile communication devices might suggest that people value this attribute the most, and, when asked, will mention it as the principal agent to eliminate communication problems arising from physical remoteness. Yet, while discourses of mobile communication devices do indeed underline how their ability to transcend space and time is profoundly transforming our perception of communication, there are suggestions that people do not necessarily experience the functioning of those devices in such bloated terms in everyday life. According to communication scholar Valerie Frissen (2000), the mobile industry's conception of information and communication technologies (ICTs) as the tools par excellence to solve communication problems is not immediately reflected in how people talk and think about those technologies. Frissen stresses that most ICTs are taken for granted or seen as double-edged swords, and therefore are not often spontaneously mentioned as possible solutions for communication problems resulting from time and coordination constraints – even when they are used as such (Frissen 2000: 72-73).

Verifying and adding to Frissen's empirical research on the role of ICTs in households, Mark Aakhus (2003) concurs that there is a mismatch between the industry's representations of improved communication and the ways in which actual uses of communication technologies are experienced. Indeed, despite the notion that a mobile telephone can provide more personal connectedness than any other medium, it is rather a stretch of the imagination to think of someone literally proclaiming that she uses a mobile telephone because it is able to solve all known communication problems. To explain the existence of such mismatches, Aakhus proposes to reframe common perspectives on communication and its problems, and to do so in two steps. The first step is to recognise that the 'pregnant image of "perpetual contact"' – which Aakhus acknowledges as 'a long-standing idealization of communication' – has a strong influence on the design, use, and thus affordances of ICTs, and therefore will continue to push increased connectedness as an ideal form of communication (Aakhus 2003: 38). In other words, user perceptions of how well ICTs can help solve communication problems can be expected to change over time as those problems become more and more articulated in terms of inadequate or faulty connections. The second step is to understand that the use of mediated communication should not solely be interpreted as a means to virtually transport to another place, but also as part of a dialectical exercise to 'resolve the competing desires and expectations to be separate or to be together' (ibid.: 40). In short, mobile telephone practices in everyday life are about solving communication problems not so much by substituting face-to-face meetings, but more by constantly reconfiguring one's connectedness in the network of communication flows.

Taking Aakhus' proposed perspective on communication as a guideline, user accounts of reconfigurations of connectedness can provide a deeper understanding of how people seek to exploit the range of communicative affordances of mobile communication devices, and subsequently of how they experience what are supposed to be idealised ideas of communication sparked by the desire for perpetual contact. To expand on this, there is one particularly poignant theme that runs through such user accounts, a theme that underscores the conclusion drawn previously that mobile telephones increase one's connectedness without necessarily diminishing face-to-face meetings. This is the theme of social coordination, or rather that what Richard Ling and Birgitte Yttri (2002) distinguish as a combination of 'micro- and hyper-coordination'. Micro-coordination, according to Ling and Yttri, is the type of nuanced instrumental coordination typical of a significant part of mobile telephone use: trips that have already started can be redirected, people can call or text to say they will be late, and meetings can be scheduled at a rather loosely defined time or location, only to become more definite when those who want to meet call each other while they are on their way

(Ling & Yttri 2002: 143-144). Hyper-coordination brings an expressive layer to this instrumental use, both in the form of social and emotional communication (chatting, gossiping), and in the form of mobile etiquettes telling where mobile telephones should not be used or which models are in fashion (ibid.: 140). Both types are specifically about how to manage increasingly connected social networks: micro-coordination in a logistical way that makes full use of enhanced communicative availability, and hyper-coordination in a cultural way that establishes the mobile telephone as today's *nec plus ultra* means to create, maintain, and express social bonds and values. In user accounts of both types, elements of ideas of fulfilment through improved communication seep through voiced expectations and desires.

First examining micro-coordination, Ling and Yttri find for instance that in many comments about the motives for using mobile telephones a strong need for connectedness is conveyed, with the goal of making instrumental communication proceed as smoothly as possible. As one informant of an interviewed group summed up reasons for contacting her partner: 'It is if somebody is late, it can be if we need to buy something, if there is something important that he needs to bring home, if he needs to call somebody or if he has been home and has to give me a message. It is not like "Hi, I am doing fine, etc." It is something that we need' (ibid.: 145). This instrumental approach to mobile telephony is widespread, if not constitutive. Ling and Yttri retrieved their data mainly from interviewing Norwegian teens, but similar attitudes towards the importance of being connected for instrumental reasons have already been registered in early social analyses of the adoption of mobile telephones (see Kopomaa 2000) and have been identified as largely age-indifferent and cross-cultural phenomena (Castells et al. 2007; Leonardi, Leonardi & Hudson 2006). In such conversations, the emphasis in communication does not lie so much on reaching mutual understanding through dialogue, but more on orchestrating each other's movements and positions in the space of flows to the point where they ultimately overlap and merge. This is only possible, however, if permanent reachability and availability is guaranteed and is incorporated in the ideology of mobile communication. As another informant assessed his need for increased connectedness: 'It is practical to be available because you do not miss anything. It is also practical that others are available; at any rate it is irritating not to be able to reach people when you want to reach them and get them involved in something. It is very irritating' (Ling & Yttri 2002: 151).

Echoing the diagnosis of the paradigm of networked communication in chapter 2, the main motivation here for using mobile communication devices lies in the assumption that if there are more opportunities to connect there will be an improvement in how we arrange social interactions, and thus in how social groups function as a whole.[14] Of course, this assumption only holds when there is an agreement on how social interactions should actually be arranged, but in the mythical mobile vision – availability solves everything, together is good – the mere possibility to contact anyone from anywhere is enough to suggest that such problems can easily be dealt with, simply by making another call or sending another text message. In the informants' justifications for using mobile telephones we see that communication is not so much actually improved, but it is *presented* as such, upholding the myth that the devices have indeed made communication better.

In hyper-coordination, the value of being connected is less related to efficient planning or dealing with practical issues, and more to achieving a certain status and maintaining intimate social bonds. Such communicative activities are about sharing experiences and confirming personal links, and can take the form of gossiping, catching up on each other's adventures, or exchanging symbolic gifts (Johnsen 2003). Now, these practices have been around for a long time, but the ubiquitous connectivity that the mobile telephone promises to deliver represents a new and particularly potent means to establish one's identity and earn a meaningful place in the social hierarchy of family members and friends. In a cross-cultural study done by Scott Campbell (2007), results showed that people find that through mobile telephones group connections are enhanced, conversations gain in intimacy, and there are more opportunities for emancipatory praxis. One informant in the Ling and Yttri study illustrates this last point: 'If I am not home and if I didn't have a mobile telephone then my parents would be clear about all the people I hang out with [...]. When you have a mobile telephone then you have a private answering machine and a private telephone' (Ling & Yttri 2002: 153).

The mobile telephone thus acts as a device that appears to liberate expressive communication, largely because it makes opportunities for mediated dialogue and dissemination available to demographic groups that did not readily enjoy that privilege before the age of mobile communication devices, and because those people use mobile communication as a way to actively establish and manage their own social position in relation to others. Here, the ideal of communication as a free, uninhibited practice resonates quite strongly, and in a literal sense as well, of course, as expressive communication is also liberated from its physical attributes. Expressing oneself as a unique individual now increasingly takes place in the virtual media space of mobile conversations, where social differences tend to be less visible or obtrusive. Ling and Yttri find that many mobile telephone users think that the most important aspect of coordination activities is that they can be performed without the involved parties knowing their respective locations, which adds a highly egalitarian and personalised quality to already existing patterns of communication (ibid: 143). Compared to the fixed telephone, the mobile telephone offers far more direct and individualised links to other people, properties that have quickly turned the device into the primary locus where one's collection of social connections resides. In the words of another informant: 'I think that the mobile telephone is most important in relation to my friendship network because I have the memory full of phone numbers and that is only mobile numbers. I could not get in touch with my friends at their home phones' (ibid.: 152). Being in touch and expressing that connected status is what matters in hyper-coordination. Because mobile telephony is perceived as extending access to others to virtually infinite dimensions, the thought of being perpetually connected in the symbolically charged mobile network gives an increased existential significance to relational ties. As Gergen put it: 'The Enlightenment paean to individualism, "I think therefore I am" is replaced with "I am linked therefore I am"' (Gergen 2003: 111).

But, as in all necessary fictions, there are distortions in the image of communication utopia; increased social connectedness comes with some unintended consequences. The fact that mere connectivity has come to engulf present-day imagery of communication has a compelling effect on our relational self, changing the perception of what it means to co-exist and communicate with others. The people we interact with most – those who reside in our more intimate social circles, like family members, friends, and colleagues – are now always only a phone call or text message away. As Gergen aptly has diagnosed this social condition, we find ourselves continuously in a state of 'absent presence': physically absent, but electronically at hand (Gergen 2002). While the arrangement of absent presence offers a psychologically reassuring feeling of closeness, perpetual connectedness does not necessarily guarantee reaching a global coming together; Gergen notes that because communication with absent present others takes place via readily available channels, and therefore does not require a lot of effort or time in setting up, it tends to become simpler, shorter, and distributed among several fragmented micro-communities (Gergen 2003: 106-107). In addition, when people are inclined to enclose themselves in connections to their absent present social network, they exhibit diminished concern for those outside of their communicative bubble (ibid.: 109).

Thus, while more and more people gain increased opportunities to connect to each other, there are reasons to believe that they tend to use mobile communication devices predominantly to maintain already existing connections, and do not necessarily employ them to expand their social network (De Gournay & Smoreda 2003). Newly acquainted people might not necessarily become new mobile connections, either because contacts are fleeting or because they do not wish to be added to one's group of existing connections. Moreover, adds Myerson, the dominant mobile imagery in advertisements and press releases even enforces the idea that the maintenance of those existing connections is more about fulfilling personal desires than it is about getting to know each other better. He states that by continuously touting the growing amount of mobile telephone users while at the same time stressing that communication is something that you personally can control, a paradox is created: 'On the one hand, we have a language of scale; on the other hand, we have the separate individual seeking goals' (Myerson 2001: 20-21). Even more so than the fixed telephone, a mobile telephone is part of a highly intricate, global communication network, all the while being a personal, individualised communication technology that favours personal, individualised contact.[15] All the necessary (but mostly technological) conditions for a completely interconnected world may be in place, but a realisation of a grand togetherness is not automatically in the cards.

What is noticeable as the mobile network grows, of course, is that mediated communication is brought out more into the open, which presents us with new scenarios for experiencing being-with-others and strengthens the notion that anyone carrying a mobile communication device can, potentially, become part of anyone else's technology-mediated network. The drawback of the invasion of the public by the private, however, is that mobile communication behaviour in public spaces often invades and disturbs social events and face-to-face conversations with what is perceived as trivial and redundant chitchat or gossip. Now, gossip is the basic social glue with which humans build and maintain social bonds, and mobile telephones are very apt at facilitating it anywhere and round-the-clock (Fox 2001), but when experienced intrusively and only one-sidedly it tends to aggravate companions, bystanders, and eavesdroppers, causing them to feel disempowered (Plant 2001: 31) or convinced that communication in the mobile age has become tawdry (Palen, Salzman & Youngs 2000: 207). This strained intermingling of public (outside) space with private (inside) space is what psychologist Kathleen Cumiskey (2005) calls the paradox of techno-intimacy: to ourselves, our mobile telephone is a highly convenient personal item and our own mobile communication behaviour is perfectly acceptable, but we tend not to appreciate the same behaviour and attitude towards the valuation of mobile communication in others. In this sense, techno-intimacy, the ambivalent relationship between the desire to be connected and know all on the one hand and the need to stand apart from the multitude on the other, is itself a typical exponent of the paradox of idealised communication: an achieved complete togetherness will necessarily entail the loss of individuality, and mobile telephony will continue to stress this phenomenon.

Real-time relief of anxiety

If we perceive mobile telephones to be personal devices full of communicative potential that, through their capability to transgress space and time, can maintain and strengthen bonds with primary social group members and make coordination activities more ad-hoc and prominent, one of the following observations must be that the psychological 'fix' attached to wireless communication behaviour heavily relies on the involvement of feelings of reassurance. Providing relief is the perfect medicine for those that desire to be free from the anxieties that necessary fictions of progress can produce, and the mobile telephone promises to do exactly that when it comes to dealing with the high expectations that stem from the hope that communication will be improved. For many, the emotional immediacy of the device has registered itself as an indispensable part of their everyday life, like the aura of a talisman worn for good luck and protection. The ability to 'just call' or 'just text' and receive a confirmation of an appointment, an answer to a question, a promise of help, or an acknowledgement of an intimate relationship, preferably as quickly as possible or at least within socially acceptable time limits, makes for a very appealing advertisement premise, as well as for a credible and legitimate motivation to own a mobile telephone.

And indeed, research has shown that a need for security, safety, and reassurance is high on the list of initial reasons why people decide to buy mobile telephones, either for themselves or for their children or other loved ones. In a cross-cultural field study undertaken in Berlin, San Francisco, Shanghai, and Tokyo, it was found that mobile telephones were, alongside keys and money, 'considered essential irrespective of culture or gender' for 'survival in the modern urban world' (Chipchase et al. 2005). In a 1999 qualitative survey of 36 focus groups in six European countries, Richard Ling reports,

> respondents were asked to what degree they agreed or disagreed with the statement 'The mobile telephone is useful in an emergency.' We found that approximately 82% of the respondents were in complete agreement. There was no other attitudinal indicator with regard to mobile telephony that had such an extreme score. (Ling 2004: 37)

Similar results were obtained from studies in the United States and Australia (ibid.: 38). In yet another study in the United States, safety and security were identified as forming a common, broad category of motivations, which were 'often associated with car-related safety or for unknown situations that might arise' (Palen, Salzman & Youngs 2000: 204). It should not come as a surprise to see the car mentioned here; much as was the case with telephones made mobile by installing them into cars, the modern day mobile telephone is perceived as an ideal means to travel and still remain in touch with the outside world, offering the comfort of knowing that when something goes wrong, help is always at hand.

The human fear of calamities and disasters evidently provides a sound reason why feelings of security and safety play a significant role in the adoption of wireless communication technologies. The many modern folk stories that tell of amazing rescue missions that involve the use of mobile telephones are adamant examples of how strong the desire is to stress that increased connectedness is a 'good trick'. Idealised ideas of communication thrive on compelling anecdotes that allegedly prove that progress has been made, and arguably nothing provides more conclusive evidence than accounts of lives saved or loved ones protected thanks to new communication technologies. So, we read in *The Guardian* that two British climbers, caught in a blizzard on a Swiss mountain, texted five friends, one of whom received the message in London at 5am and immediately notified the rescue services in Geneva. After having waited another 36 hours because the conditions were too severe for rescue teams to pick them up, the two climbers were finally saved (Allison 2003). Similar stories tell of people having become lost in a pass, shipwrecked on a boat off the coast of Indonesia, or stranded in the outback of Australia, who were all able to alert friends or family through their mobile telephones and consequently receive life-saving help (Turrettini 2004). In the same vein, when in desolated urban environments people find themselves in threatening situations or feel they are intimidated by the presence of strangers, mobile telephones can sometimes offer a sense of protection just by their powerful symbolism of connectedness (Ling 2004: 44-45). Such stories thus readily feed myths of improved communication: without the new mobile communication technologies, lives would have been severely impaired, or even lost.

However, increased security through these technologies can only be guaranteed if people are willing to sacrifice some or even all control over when they can and should be reached, and by whom. Forceful evidence of this highly charged problem created by the need to relieve anxiety can be found in one of the more common social relationships where reassurance plays an important role, namely that between parents and adolescent children. On the one hand, the mobile telephone offers parents the ability to let their children discover the world on their own, with the added safety of knowing that the teenagers can always call in case of an emergency. As one Norwegian mother in a group interview acknowledges:

> I have a 17 year old and the worst thing I know is when she goes downtown. I am so afraid but I just have to accept this you know. But it helps that she has a mobile telephone because she can call if something happens. It is not to control my daughter that she should take her mobile when she goes out, but it is, [...] 'If something happens, call home and we will come immediately!' you know. (Ling & Helmersen 2000: 14)

Here we see the familiar themes of safety and emergency closely associated with the mobile telephone, emphasising its function as a communicative lifeline that can be used at any time and place.[16] In this sense, it shows its appeal as an idealised medium: it is the ever-present 'materialiser' of fulfilment of hope, hope that everything is and will remain fine. Yet, on the other hand, even though we hear the mother say that she does not want to control her daughter, notions of surveillance and accountability are exactly what will follow from a heightened absent presence of others. When one's whereabouts and activities are continuously under potential scrutiny, the mobile telephone becomes a mobile leash, exerting a strong influence over its carrier.[17] In their adolescent quest for independence and in response to over-concerned parents, as sociologist Nicola Green has found, teenagers often develop 'parent management strategies' with which they regulate their reachability (Green 2002: 39). Some say they had not heard the device ringing, others automatically redirect their parents' calls to voice mail, and there are those that just turn off their mobile telephone when they go out, and say that their battery was dead.[18]

While Green notes that what exactly constitutes surveillance and accountability in mobile relationships is contextually based, and that the dominant association of these concepts with state-controlled law enforcement does not do justice to the new ways in which individuals gather and share information, what should be regarded as most important in her analysis is her contention that the proliferation of wireless communication technologies has 'normalised' the activity of checking up on others (ibid.: 33). Through mobile telephones, Green writes, individuals 'engage in routine monitoring of themselves and each other [...], and assume that others are self-regulating and accountable for their use of devices in both co-present and tele-present contexts' (ibid.: 43). Thus, everyday notions of what it means to feel reassured increasingly come to rely on knowing what others are up to, because the technological means to gain that knowledge are at one's disposal at all times. As a result, questions of privacy become more manifest.[19] As sociologist James Rule contends, compared to older mass media systems, today's infrastructures of perpetual contact inherently generate more personalised information available to large institutions, corporations, and groups of individuals, making a future world of 'total surveillance' a conceivable reality (Rule 2002: 247). Even though, like Green, Rule does not want to attach a specific value connotation to the term 'surveillance', he does point out that we should be cautious not to easily dismiss the dangers of this outlook, or think that we can always escape observation by simply turning off our connections when we want to (ibid.: 248). In the mobile age, perpetual contact becomes the norm, and participating without it difficult indeed.

These observations of ambivalence in mobile reassurance technologies highlight the pitfall of the desire for unlimited communication: the more opportunities are created to connect and communicate, the more the struggle will be to hold on to established boundaries between the private and the public, between what can be known, should be known, and needs to be known. Mobile communication devices may be able to function as symbolic crowbars, breaking open social patterns of communicative behaviour, but they can only do so at the expense of disclosing a lot more information than people might care to consider. What is more, even if the aforementioned boundaries would blur to such an extent that more data would become freely available than otherwise possible, there would still remain the nagging uncertainty whether that retrieved information is truly enough for one to be relieved of all anxiety. In fact, by seeking reassurance through radical connectedness, another type of anxiety is created, one that stems from simultaneously being connected to someone far away and knowing that that distance cannot be bridged physically. This spatial discrepancy may not be so problematic when a call is made for trivial reasons, but when related to an emergency, the state of being distant but present can instill overwhelming feelings of isolation and powerlessness.[20] A fickle balance between fear and relief manifests itself in the continuous search for reassurance; when Henrietta Thompson, for example, states in *Phone book* that '[mobile] phones offer the best peace of mind it is possible to get', she does so in the context of Israel's continuous preparedness for terrorist attacks, where '[with] the fear of [terrorism] always present, the need to communicate is paramount', and 'people need to be able to check-up on their loved ones on short notice' (Thompson 2005: 55).

So, not only does the longing for perpetual contact create problems with privacy issues, it also multiplies instances of what we could denote as 'terrifying closeness', moments where mobile communication enables people to be connected in extreme emotional circumstances while being physically apart. The torment of such 'intense immediacy', as James E. Katz (2006: 104) calls it, becomes adamantly clear in situations of life and death, because, while wanting to let loved ones know they are on the mind of a dying person is an understandable human emotion, to think that mobile telephones merely extend the possibility to do so is to woefully misapprehend the highly disturbing perception of experiencing remote deaths. Imagine, for instance, the unimaginable conversation between the stranded mountaineer and his wife, whom he had called from somewhere on Mount Everest to say he was going to die (Cusk 2001). Or think of those other, highly profiled accounts of mobile calls in which Eros met Thanatos, made during and after the terrorist attacks on the World Trade Center in New York on 11 September 2001. People in the hijacked planes and those trapped inside the stricken towers called family members to tell them that they loved them, sometimes right up to the moment when they died. After the towers had collapsed, rescue workers could hear people from under the wreckage use their mobile telephones to call for help, not seldom in vain. Most dreadful are the accounts of unanswered mobile telephones ringing in the rubble, or even in body bags.[21] All these cases vividly illustrate how feelings of reassurance and distress can come palpably close to merging into a single sentiment, how wireless communication technologies can augment both intimacy and isolation to such a degree that unlimited communication is indeed virtually reached, in all its real-time glory and ugliness.

Omniscience and collectiveness through ever-present knowledge

The final characteristic of wireless communication technologies examined here is, like the three discussed above, intimately connected to topoical expectations of what new communication technologies ideally should be capable of. It is the project of freeing access to information and knowledge, which, when pressed to its radical ending point, should enable anyone to know anything on whatever topic, and become a member of a completely transparent society. This is a familiar utopian narrative, of course, and the mobile communication age has sparked renewed belief in necessary fictions that tell of ways to make this outlook a reality. Personal wireless communication devices such as mobile telephones are presented as enabling technologies with emancipating powers, giving instant and ubiquitous access to people and information resources which would not have been as easily – if at all – available in the days before wireless communication technologies. The emphasis in such imagery is often on reaching harmony and agreement through the exchange of knowledge, and on making progress through the fusion of ideas. The questions are, though, if these communication technologies are indeed experienced by their users as such, and to what extent idealised ideas of making all knowledge and information accessible clash with the realities of everyday mobile communication.

The force behind the current knowledge and information paradigm in mobile discourse is 'largely technology-driven' (Castells et al. 2007: 110). One thing that is readily noticeable in the evolution of mobile telephones is that, since the 1980s, there has been a growing complexity of communication possibilities, which offer more and diverse ways to interact. First, obviously, there was calling. This facilitated spoken dialogue between two individuals who could be located anywhere, and, especially in its early business-oriented days, it advanced the notion that people themselves had become more available as social and informational resources. Then, with the development of the GSM standard in the early 1990s, text messaging (or short message service, SMS) was added as an asynchronous 'store-and-forward' service (Goggin 2006: 69). Although since the end of the 1990s SMS has been appropriated by mainly young people as a means to keep in touch, its initial proposed purpose was to function as a 'unidirectional system for sending "mobile terminated" messages' (Taylor & Vincent 2005: 79). Both types of SMS use have been retained over the years, and nowadays SMS enables individuals not only to engage in mediated dialogue, but also to disseminate messages among larger groups of people, receive automated notifications, and send codes to computers which are then processed by specialised software. So, at the beginning of the 2000s, mobile telephones did more than just handle telephone calls; they had been integrated into a diverse range of social practices that involved retrieving and transmitting knowledge and information. Then, over the last few years, bandwidth capabilities were drastically increased and additional features were packed into mobile telephones: photo cameras, music players, games, and Internet browsers and apps all found their place in the small technological wonders, transforming them into smartphones.

We can see that entertainment and data processing functions have increasingly become significant components of mobile communication. As Castells et al. put it, '[m]obile communication devices are the multipurpose, multi-channel connecting points in the network of communication of which everybody becomes a personal node' (2007: 110). Their description of people as 'personal nodes' here is telling, as it raises the question whether entertainment and data processing somehow corrupt the social nature of the mobile telephone by transforming it into a converged set of individualising technologies.[22] However, while the process of adding new features has largely been the result of the mobile industry's search for more revenue, the quick acceptance by the general public of these features also testifies to the growing cultural importance for the individual to be connected to her social network in a variety of ways, using more than just speech and text to communicate. Moreover, both industry push and consumer pull at least in part originate from the same realm of necessary fictions that say that increased and diversified connection possibilities are what improve the sharing of knowledge and the advancement of human cooperation. So, the individual may be the preferred target of selling strategies, but, crucially, the heterogeneous interlinking of personal nodes has also intensified the awareness that knowledge and information become more decentralised in the mobile age, making all connected individuals part of an ever-present collective network where every member potentially is in the know. This awareness, which clearly echoes ideas of idealised communication that were found in perceptions of networked communication in chapter 2, can especially be located in present-day discourses of the mobile/wireless Internet and of the perceived democratising nature of mobile telephones.

The Internet carries with it, as we saw earlier, the enlightened ideology of universal access to all available knowledge. As a medium of media and the facilitator for the global addition, storage, transmission, and retrieval of whatever kind of information by anyone with a connection, it remediates all previously known communication technologies, and as such it has inherited the hopeful aspirations of every expectation expressed so far that a communication utopia is finally within reach. Tragically so, it turns out, because there is always the doubt that we are not quite there yet, that there is still something that needs to be improved. The current perceived problem of the Internet, aptly voiced by media scholar Paul Levinson, is that the personal desktop computers that give access to the Internet make us 'highly dependent on walls in rooms': they require lots of energy, and thus are in need of electrical outlets (Levinson 2004: 38). The solution, he says, lies in wireless communication technologies, which can bring the Internet out in the open to create a world of 'immediate information' (ibid.: 55). With his 'remedial' approach to the evolution of media, in which he sees developments as direct reactions to perceived disadvantages of existing media, Levinson here conveys the charged yet familiar and mythical notion that there is a 'natural desire' for communication to be free and on the move, that making the Internet wirelessly accessible satisfies a human longing (ibid.: 9-15). And indeed, we see that mobile technology industries treat the development pattern of wireless Internet as similar to earlier, established 'good tricks' in media evolution, because it, as one enthusiastic chronicler of wireless Internet writes, 'covers all communication needs of human beings' (Jamalipour 2003: 3).[23]

Such sweeping and skewed remarks are not uncommon in discourses on wireless Internet. The perceived magic of wireless technologies, combined with the imaginative power of a global information system, quite palpably creates a language of communication that harbours very familiar idealised statements about improvement, fulfilment, completeness, cooperation, and so forth. The aptly named British wireless broadband provider The Cloud, for instance, promises on its website to deliver '[p]*roper* access to information, communication tools and mission-critical applications', which will advance productivity by increasing work effectivity and 'keeping your people connected away from home or the office' (The Cloud 2007, emphasis added). A similar, but user-operated and not-for-profit project is run by Community Wireless, an organisation that, when advocating the sharing of wireless Internet connections with community members, says it represents 'a global dream', and propagates on its website the coming of 'the Organic Internet. The Internet re-born. The Internet the way it should be' (Community Wireless 2001). Such utterances are not limited to advertisements or idealists' rhetoric, however; analyses of end users' perceptions of wireless Internet applications have also revealed that people expect and experience advantages in terms of 'newer and better ways of being [...] in touch with information' (Ng-Krülle et al. 2004: 5).

With the advent of mobile telephones that support high-speed data transmissions, as well as with the growing number of Internet-enabled devices such as iPhones and Symbian/Android-based smartphones, wireless Internet increasingly finds its way into the palms of many hands.[24] In the past five years mobile telephone producers and service providers have started stressing that the current crop of devices represent that 'what computers have become' (Nokia 2006), that they finally connect you in any way to 'the people who matter most' (T-Mobile 2010), and that life will become better as '[f]rom now on, we all have more time, because [...] the Internet is now truly mobile, so you can make use of every minute of every day' (Vodafone UK 2007). All these slogans of course serve to point out that the marriage between the Internet and mobile communication devices is a successful match, but they also subtly push the idea that the arrival of wireless Internet is part of a natural process, one that is on a trajectory of fusing all information and people together in a world where opportunities to make use of those fusions are abundant.

The second important area in which we can identify the idealised notion that, through their attributed inherent ability to share knowledge and information, mobile communication devices are instrumental in the construction of an egalitarian space of collective intelligence, is in accounts of their supposed democratic nature. Especially mobile telephones have become powerful symbols of both communicative liberation and social inclusion, and as such are seen as the instruments of choice for bridging all forms of divides, be it the political divide, the economic divide, or any other demographic digital divide variable. The main reason for this symbolic investment lies in the perception of the role of earlier mass media in constructing a homogenised and even imposing system, in which dominant power relations and an undemocratic control over communication and information networks largely prohibited individual expressions of opinion to be made public. Visibly making use of myths of the technological sublime, criticism of the mass media has taken the shape of celebrating new, decentralised, and personalised communication technologies, with which old communication obstacles are thought to be eradicated. Thus, next to the Internet, mobile telephones are often touted as a very effective means to circumnavigate the disadvantages posed by the mass-media system, and to create new networks of cooperation. Imaginative stories of how they enable people to group together in what Howard Rheingold (2002) calls 'smart mobs', and subsequently oust presidents, overthrow governments, or orchestrate massive ad-hoc demonstrations, all serve as powerful new chapters in necessary fictions of hope and progress achieved through improved communication technologies.

While such stories are prone to overestimate the role of mobile telephones and are often more anecdotal than substantive, they do convey a demonstrable and widely held understanding of wireless communication technologies as catalysts of changes in societal power relations. With the caveat that he does not want to suggest any inevitability about mobile telephones opposing centralised power, science and technology historian Jon Agar, for instance, sees 'a correlation, a sympathetic alignment, between the mobile phone and the horizontal networks that have grown in the last few decades in comparison with older, more hierarchical, more centralised models of organisation' (Agar 2003: 162). Similarly, after having been equally careful in avoiding the construction of a technological determinist argument, Castells et al. assert:

> Still, it cannot be denied, based on the observation of recent processes of sociopolitical change, that access to and use of wireless communication technology adds a fundamental tool to the arsenal of those who seek to influence politics and the political process without being constrained by the powers that be. (Castells et al. 2007: 212)

Other empirically gathered evidence of how mobile telephones can assist in throwing off the shackles of poor government and economic disadvantages can particularly be found in developing countries, where micro-entrepreneurs use the devices to access price information themselves instead of relying on suspect state-controlled media reports (Donner 2003). So, even though we should not discount other factors that play a part in social, economic, and political change, we can say that wireless communication technologies are perceived as adding a significant impetus to bringing about that change. The idea, again, is that by unlocking channels to information and knowledge, they empower the individual.

We should be careful, however, not to identify the empowerment of the individual and the growth of horizontal networks as unique utterances of the idealised desire to create harmony and ultimate togetherness through improving communication. Nor would it be wise to ignore the drawbacks of these processes, which define just as much how we experience the change towards the mobile communication age. As more and more people are connected, the kinks in idealised ideas of communication will become increasingly articulated. In the grand project of raising access to knowledge and information to ubiquitous levels, one of the more conspicuous problems for instance is that, as the flow of information grows over time, more doubts will arise about the value of retrieved information, as there will always be more to evaluate. As James E. Katz notes, 'technologies of freedom' such as the Internet and mobile telephones aim to establish 'pluralism of expression rather than a dissemination of prefabricated ideas', and as such they create ample room for struggles over meanings and interpretations (Katz 2006: 151). Thus, in what he sees as an ironic reversal of Claude Shannon's axiom that information is uncertainty reduction, Katz holds that 'increased information also leads to increased uncertainty' (ibid.: 152). A similar observation comes from Joshua Meyrowitz, who, after having explored how wireless communication media create new possibilities for 'non-geographic "groups" to act together', notes:

> Ironically, the increased potential to access, juxtapose, compare and contrast, and construct alternative narratives is often paired with the reduction in the psychological inclination to engage in such time-consuming analysis. Paradoxically, the more our new technologies allow us to accomplish in an instant, the more we seem to run out of time. (Meyrowitz 2003: 100-101, emphasis in original)

Meyrowitz here reaffirms the notion that the ever-returning hopes and promises that social harmony will automatically increase with improved communication continuously clash with the reality of everyday communication, in which the need to filter and select is strong but the ability to do so is limited. Echoing the problematic effects of the desire to achieve communication utopia discussed earlier, such clashes in the swelling streams of information can induce anxiety over reachability, transparency, and accountability. What information is true or useful, and what is not? Where did it originate, and from whom? Just as mobile telephones can easily aid in spreading important news or calls for action, they can also multiply rumours and inaccurate information, perhaps even more so than other media because of their personal nature (Fox 2001; Castells et al. 2007: 212). Furthermore, instrumental as they may be perceived in the possibility of connecting people in a truly democratic fashion, their use is predominantly geared towards maintaining existing social networks, and not towards integrating into a larger social cohesion (Fortunati 2003: 249-250). Again, this reveals the contradictory human tendency discussed earlier to also *resist* the implications of the formation of a 'grand togetherness', and to be wary of a state of existence in which individuality and belonging no longer carry their current meanings.

As was the case with the previously discussed characteristics of mobile communication, by radically opening up access to and production of knowledge and information, wireless communication technologies fervently confront us with the paradoxes of pure communication, getting us closer to the fulfilment of the desire for unlimited connectedness but at the same time letting us experience its ruthless blending of all actors involved. This process will continue to go on, co-oriented by the myths that say that there is still room for more improvement.

Improving perfection

> *Technology. Making 'better' better. Onwards, upwards, any way but backwards. Tapping progress on the shoulder, and saying: more forwards please!* (Honda 2006)

If there is one thing that should be adamantly clear after the analysis of how idealised ideas of communication find their expression in wireless communication technologies, it is that new devices may on the one hand be presented and experienced as truly bringing important prerequisites of pure communication a big step closer to becoming a reality, but that they, on the other hand, strikingly resemble other and older media technologies in being just another tragic attempt in the never-ending quest to reach communication utopia. As with those earlier attempts, both wilful amnesia and myths of progress have accompanied wireless communication dispositifs from the onset, making sure it is quickly forgotten that the media we have now are themselves already the result of trying to improve things, and telling us we need to improve once more. Thus, mobile communication devices are unstable media, always part of an evolutionary process that knows no end, always susceptible to all kinds of social, political, and economic factors that, together with idealised ideas of communication, lay bare strong as well as weak spots and open up new paths for improvement. Because we continuously run into the paradoxes of the desire for pure communication, we keep on eliminating as well as creating obstacles, and therefore set ourselves up for the inevitable lure of the beckoning future or for the looming threat of the onrushing future. There is still more road to travel and still more to add to what we have got, is the prevalent attitude. Today, there are two widespread developments in the wireless communication paradigm that are presented or perceived as making mobile communication devices even better than they are now: the introduction of location awareness capabilities and the creation of the Internet of Things.

Location awareness

One of the most noticeable additions to mobile communication devices in the past few years has been the possibility to have them 'know' where they are in geographical space, and thus make them location-aware in their functioning. Of course, because cellular technology works with uniquely identifiable base stations that provide radio coverage for mobile telephones in small overlapping areas (or cells), a certain knowledge of where the devices are located has always already been part of the cellular system. However, the range of these cells can vary between seven hundred metres to as much as 70 kilometres, making it very hard to pinpoint exactly where a connected mobile device is when only its cell identification is used as a marker.[25] It is only now, when GPS processors are slowly but surely becoming a default feature of mobile telephones, that their location can be determined much more precisely, with a typical accuracy that ranges between less than a metre and about fifteen metres. Fuelled not only by the mobile industry's need to create additional streams of revenue, but also by FCC and EU regulatory decisions aimed at aiding rescue workers in accurately responding to emergency calls made on mobile telephones, GPS-equipped communication devices are quickly becoming commonplace.[26]

The effect of this beefing up of wireless communication technologies promises to be profound. While communication devices such as mobile telephones first transferred mediated conversations from the 'situatedness' of fixed connections to the highly nondescriptive 'anywhere' of media space, their location awareness now reinserts a spatial variable into our understanding of mobile communication. Not only does this potentially reconstruct earlier conceptions of what it means to be spatially connected, but it also incorporates location as yet another type of accessible information into our growing sense of omniscience. From this process a new kind of hybrid space emerges, one where additional information layers have been added to physical locations, and where a user of a mobile device will have been transformed into an even more tightly integrated node in an ever-expanding information network (Vries 2012). In such a hybrid space, more and more variables will be stored in databases and become available for query, to be used in many conceivable social situations or emergencies (see Gordon & de Souza e Silva 2011). The digitisation and mobilisation of location can thus be perceived as adding yet more weight to the persistent idea that mobile wireless communication technologies can (and are expected to) offer us access to any type of information, anytime, anywhere.

It is therefore hardly surprising that location awareness is researched and marketed as vigorously as it is today. It presents the opportunity to proclaim renewed hope in a better, even more beckoning future where communication has yet again been improved. In a familiar display of how idealised ideas of communication come to be expressed in wireless technology discourse, the notion is put forward in press releases and technical papers alike that with location-aware devices a 'qualitative leap' is made in mobile communication, one that will 'deliver relevant, timely, and engaging content and information', and 'can *help reduce confusion*' (Rao & Minakakis 2003: 61, emphasis added). In scenarios of today's communication needs, the modern citizen is portrayed as being in danger of becoming unsettled, either by daunting tsunamis of information or by her lack of knowledge of her immediate surroundings, and, as usual, she is promised that new communication technologies will help solve those problems and make life easier, this time by interacting with her surroundings in all kinds of 'intelligent' ways. The problem-solving characteristics of new location-aware mobile technologies are especially highlighted in some of the more recent NTT DoCoMo promotional videos, which present us with a mobile life in the near future that, through a pursuit of 'Smart Innovation', is strongly integrated with positioning technologies (NTT DoCoMo 2010). There seems to be no limit to what those interactions might be: proposed applications that make use of added spatial intelligence include the abilities to navigate unknown roads, find nearby friends, locate restaurants or other businesses, receive offers from stores while passing them, play location-based games, walk 'digitally enhanced' touristic routes, and so forth.[27] The only thing that is needed to never get lost again in the myriad of data, or so it appears, is the new, context-sensitive wireless communication device.

In the rhetoric of advertisements and industry forecasts, then, location-aware mobile devices are often denoted as constituting a new breed of 'smart' technologies. This not only suggests that previous versions of mobile devices and other older communication technologies were 'dumb' and were in dire need of improvement, but also that the new devices have gained in autonomous behaviour. The rationale behind the use of 'smart' as an ideograph here fits perfectly with the discursive strategies that idealised ideas of communication would typically produce: with added intelligence at hand, packed in a small technological marvel, the burden to communicate without obstacles cannot be but alleviated, so it seems. As a result, the perception of what is being 'smartened' is not restricted to the devices only; in one smooth sweep, communication itself also becomes smart, intelligent, efficient, and freed from obstacles. Again, the fact that the paradoxical consequences and dilemmas of an actual fulfilment of such a smart future are cunningly occluded illustrates not only how promises of a better future depend on tempting images in general, but particularly on the portrayal of communication as something that through its improvement has helped us getting to that better future in the first place.

The Internet of Things

From the mobile world in which handheld, networked, and location-aware devices can interact with their surroundings and create an additional informational relationship between people and their spatial context, it is a small step to envision a world where such 'intelligent' connections have become even more pervasive, and physical objects themselves can compute and communicate information across wireless networks. This is more than just a science fiction fantasy; it is a prominent line of thinking that is visible in discourses produced by the wireless industry, and already an actual fact in personal micro-networking environments and in domains where logistics operations and tracking goods are of particular importance. Several enabling technologies have been developed in the past few years to make the ubiquitous computing scenario a reality, of which Bluetooth and radio-frequency identification (RFID) tags are the most common. The Bluetooth standard, aptly named after the medieval Danish king Harald Blåtand for his unifying powers,[28] was created to make multiple devices communicate with each other through the use of radio transceiver microchips. Such Bluetooth-enabled devices can detect and connect to each other to form small data networks, in theory linking up very heterogeneous technologies such as mobile telephones, refrigerators, stereo sets, personal computers, and microwave ovens. RFID tags are somewhat comparable to radiographic bar codes, but with the added value of being able to store all kinds of contextual information for the objects they are attached to. They are very small, cheaply produced and easily distributed radio transceivers that can be embedded in items such as groceries, clothes, money notes, passports, animals, cars, or mobile telephones. Combined with growing wireless Internet connection possibilities and the integration of GPS in mobile communication devices, Bluetooth and RFID add to an ever-finer mesh of network technologies.

The discursive logic that accompanies the advent of these tiny radio transceivers provides an exceptionally fine case in point to illustrate how wireless networking technologies are readily endowed with all kinds of interconnective and unifying powers, and are perceived as the next step in bringing about an even more seamless world experience. Originally coined by former Proctor & Gamble brand manager Kevin Ashton when formulating an RFID-based solution for the logistical problems of efficiently storing and transporting his company's products, the phrase 'Internet of Things' was at first a rather bloated way of describing how, as he put it, goods could be tracked 'from manufacturing to the consumer and even through recycling' (Roberti 2002). Indeed, if we look at the predominant way in which RFID tags are currently being used, we should still be inclined to think of the term Internet of Things as being somewhat overwrought. At present, the miniature radio sensors mostly have been deployed in areas where they can provide an economic or security advantage, for instance by eliminating the need to manually scan or register bar codes on products, by reducing fraud without having to physically count and check the authenticity of chips on a casino poker table, or by increasing the amount of control over who has physical access to various infrastructural facilities (Gilbert 2005; Kürschner et al. 2010; Thiesse et al. 2009).

However, this instrumental use is not seldom seen as the starting point of what could, in a very near future, amount to the more transcendental marriage of bits and atoms, the vision of a global multipurpose network of sentient and data-collecting objects that is foreseen and popularised by people like Nicholas Negroponte, Howard Rheingold, and Bruce Sterling. Especially Sterling is of interest here; as a science fiction writer with an avid interest in all things associated with imagined and dead media, and as such a frequently invited keynote speaker on everything related to the possible futures of new technologies, Sterling has a well-informed view of some of the more radical implications of implementing small and cheap transceiver technologies on a grand scale. In his book *Shaping things* (2005), he argues that a massive deployment of RFID tags – which he, in an attempt to quench the predominantly corporate business-like language connoted by the acronym, proposes to call 'arphids' – presents us with a mode of living in which everything has become an information object that is identifiable, traceable, searchable, readable, filterable, and, most of all, *available* for use (Sterling 2005: 85-91). Subsequently, in the resulting – now more aptly named – 'Internet of Things', Sterling holds, 'many previously knotty problems simply vaporize, they become trivial' (ibid.: 93). Clearly, the notion of an Internet of Things thus comes to stand for yet another all-encompassing technological solution, one that operates as an omnipresent information and knowledge safety net, as a mental cushion that is seamlessly integrated into everyday life.

Sterling's views on the possible outcomes and projected uses of such a grand web of interconnected radiochip-equipped objects resonate across a wide range of visions on the future of communication technologies, and recall well known utopian discourses.[29] Most notably, they can be located in the 2005 annual report of the International Telecommunication Union (ITU), the specialised United Nations agency that, since 1865, recommends standards and formulates regulations aimed at strengthening international telecommunications consensus. Again, idealised ideas of improved communication can be distinguished in this influential report: the prevalent motive behind the formation of an omnipresent constellation of networked humans and intelligent devices and objects is that it fulfils the desire to, in the words of analysts from the Strategy and Policy Unit of the ITU, take the '*next logical step* in this technological revolution [of] connecting people anytime, anywhere', and create an 'Internet of Things [that] will enable forms of collaboration and communication between people and things, and between things themselves, hitherto unknown and unimagined' (ITU 2005a, emphasis added). The crux is, of course, that the forms of collaboration and communication that will be enabled *are* already known and imagined, as they will be built on deeply ingrained and very familiar desires for ideal communication that have oriented many of the 'logical' steps in media evolution.[30] The vision of the future of ubiquitous computing and communication simply takes the 'anytime, anyplace, anyone' mantra of the current mobile communication condition, and radically extends it into 'anytime, anyplace, any*thing*', stretching idealised ideas of communication even more towards their theoretical limits. Equally visible are notions of beckoning and onrushing futures, which encourage and spur us on to act upon challenges and concerns. According to the report, the enabling technologies of the Internet of Things will, especially in the developing world, offer 'medical diagnosis and treatment, cleaner water, improved sanitation, energy production, the export of commodities and food security'; if we do not choose to employ intelligent technologies that can warn us of a variety of hazards, however, we risk an increasing 'loss of life due to natural disasters' (ITU 2005b: 10-11). The future thus looks bright, but only if we abide by the presumed logical steps of technological revolutions; while the ITU is quick to note that the scenario of linking everything to everything might engender concerns over privacy issues, it sees these concerns as obstacles that can and must be overcome, and warns that if we are not aware of the future benefits, 'the development of the Internet of Things will be hampered if not prevented' (ibid.: 9). Crucially, the ever-growing dependency on large and complex technological systems is de-presented, masking the inevitable problems that come with that dependency and thus upholding myths of improvement.

In the end, the unabashed enthusiasm for what an Internet of Things holds in store for us is a continuation of earlier dreams of how communication technologies are able to bring us closer to finally bridging the gaps and removing all obstacles on the way to the communication sublime. Today, the dream is to use ubiquitous and pervasive wireless technologies to create an informational cloud, to facilitate access to an ambient intelligence, to expel anxiety over not being able to find, know, or understand someone or something, in other words, to construct a seamless information society where all questions can be answered by, in the words of the director of the Center for Bits and Atoms, Neil Gershenfeld, 'embedding the means to solve problems in the things around us' (Gershenfeld 1999: 10). When, in the future, we are all to dissolve into this global data stream, we might become like digital angels... but our wilful amnesia will have guaranteed that we will not consciously realise it.

Epilogue

It would be practical if for everyone there was something like a dictionary, in which you could look up the correct meaning if you do not understand someone. It would take away a lot of confusion. Everything would be clear and certain. But for now, I will have to keep communicating, in all possible ways and with all kinds of people, so that I will find out everything myself. And maybe, there will come a day when I will directly understand everything and anyone. And they understand me.
(We all communicate. But do we understand each other? *Anne Geelen 2006*)

In 2006, the 14-year-old Dutch high school student Serena Croes asked Anne Geelen, a television director of children's programmes, if she would be interested in making a documentary about the many ways in which people communicate these days, and whether all the various means of communication such as email, text messaging, chat software, notes, and telephone calls actually helped in improving our understanding of each other. Geelen agreed, and for a few days she followed Serena with her camera, recording conversations with friends and parents, and interviewing Serena at home. As it turned out, Serena, as adolescents are prone to do, struggled with the fact that with some people it is easier to communicate than with others, and that it can be difficult to determine, and switch between, the many different modes of communication needed to relate to her family and friends. At the end of the documentary, we see her writing in her diary while hearing her say the words I used as the motto for this epilogue: what if everyone had a dictionary, and we could look up what people actually meant when we do not understand them? Wouldn't that be convenient? Wouldn't it be even better if, someday, we could understand everyone and everything directly?

Serena reiterates the central thesis of this book in just a few sentences: at one point or another, we can or will be aware of the desire to know pure communication, and maybe secretly long for it ourselves, but the reality of everyday life leads us to realise that it will always be just that: a wish for something that will never fully materialise but will haunt us nevertheless. Serena is not alone in expressing this urgent yet ultimately unfulfillable hope for communication utopia. It is a fundamental human need to recount myths that tell of progress, of paths that lead to a utopian sublime state. These myths, powerful as they are, recurrently pervade every imaginable narrative space, constantly affirming and (re)constructing stories that establish a two-pole system: on the one end, we live in a world in which we are confronted with mysterious and age-old questions about the purpose of our existence, and on the other end is the place where those questions are answered, made redundant, or are even deemed irrelevant. The path between those poles is often presented as one of great hardship and is never really completed, but the crux of the myths is that at least there appears to be a road that can be travelled. No matter how unreal or irrational it may seem to actually arrive at the other end of the two-pole system, necessary fictions tell us it can be done.

Ideas of purpose and progress are thus employed as a heuristic tool to structure an otherwise chaotic world of phenomena and objects. The creation, telling, and retelling of stories of hope, progress, and accomplished utopias through time bolsters our belief that the promise of the new brings with it an endless range of possibilities, and that it therefore justifies any means necessary to invade this space of possibility in order to come closer to a certain ideal. From Greek myths via More's *Utopia* to Enlightenment ideals and commercial advertisements in the information age, we can see that the results of attempts to reach or construct the ideal state have always been ambiguous and never final. Invariably, all outcomes have at best shifted the status quo, leaving wide open new possibilities and maintaining the lure of the not-yet-become. The line of thinking exposed here is that there is always a tomorrow where the problems and anxieties of today could be solved; there is always a motive for foreseeing that there will be closure.

This explains the seemingly trans-historical character of necessary fictions that tell of reaching the communication sublime: full closure is never achieved, as we perpetually believe that it is just around the corner. The remarkable thing is that desiring this closure is actually a tragic feat: communicative noise is a constitutive element that cannot be circumvented, so while it is seen as an obstacle to true understanding and therefore has to be eliminated, it also makes communication what it is, namely a process that through continuous negotiations and struggles with misunderstandings defines us as unique individuals. The desire for closure by constructing the perfect language or by developing the ultimate communication technology is thus a desire to transform the human self, to have it become subjected to universalising tendencies that seek to create a world in which everyone is equal. A fulfilment of this desire would present us with the paradox of a fulfilled utopia, a sublime state where conflicts are amended, differences have disappeared, and only a zombie-like blissful existence remains. Still, even though dystopian and anti-utopian narratives point us to the sombre outlooks of worlds in which universalising tendencies lead to the totalitarian regimes that make possible the elimination of obstacles and differences, the longing for improvement towards a better, final goal abides.

This longing interacts with our continuous struggles to understand ourselves and others, as the example of Serena's wish very aptly shows. Communication ideals are predominantly expressed in terms of guaranteeing closeness and reciprocity, reaching consensus, unifying channels, making and sharing multiple connections, removing semantic fog, tuning to the right frequency, and synchronising thoughts. The ways in which these ideals are to be reached focus largely on language and technology; the reasoning being that a common universal language would remove the problem of interpretation, and ubiquitous communication technology would guarantee perpetual contact, with which people would be in direct and constant touch. We keep being confronted and lured by these myths of the communication sublime for three reasons: first, the instrumental dimension of communication, in which error-free transfer of information is emphasised, provides powerful marketing slogans that readily tap into desires for improvement; second, our notions of what lies ahead of us and what we have to do to get there are influenced by images of beckoning or onrushing futures, and both these images regard communicative noise as an obstacle that either is or has to be removed; and third, future-making in itself generally presents only clear-cut outcomes, and conveniently leaves out the noisy consequences of the programmes that are presented as leading towards those outcomes. In other words, there is a clear symbiosis between imagining what improved communication could be, and acting upon it: on the one side, those working in the business of producing and selling new communication technologies make very good use of our susceptibility to myths about ideal communication, and on the other, we let them do so, as, in the end, we think the same.

We should therefore be fully aware of the strong influence of myths in orienting inventors, scientists, entrepreneurs, producers, marketeers, politicians, consumers, and so forth, towards creating specific materialisations and uses of communication technologies. The interpretative flexibility of media may engender many media dispositifs, but there are limits to that flexibility, and an inquiry into myths of communication can help identify what those limits are. In the words of philosopher Lee Worth Bailey, '[a]n entirely new consciousness is needed', one that is aware that 'technological culture is teeming with dreams, visions, hopes, goals, expectations, and imaginative premises', and that we need to understand that technological thinking 'contains a far larger and more influential component of enchantment by unconscious fantasies than is commonly acknowledged' (Bailey 2005: 17). This is exactly what a media-archaeological and evolutionary approach to technology development can provide: it has a non-teleological view on history and it acknowledges the various environments in which military, economic, political, scientific, and social factors play a role in shaping technologies, but it also recognises the agency of necessary fictions that orient development towards 'good tricks', or, in other words, towards the material expressions of our desires for utopian communication. When we look at media history from this media-archaeological and evolutionary perspective, we see that, despite reality proving many utopian claims wrong, new communication technologies have been and still are presented and perceived as providing opportunities for finally realising true democracy, cultural and social unification, the unlocking of all the world's knowledge, and so forth. It is what 'the new' does: it perpetually gives our technological imaginary, our yearning for wholeness and completeness that is projected upon technology, fresh impulses by portraying existing technologies as inadequate, and, in the same sweep, by introducing us to the next big thing as a solution.

Mobile communication devices are undoubtedly the next big thing of today. They did not suddenly originate in contextual isolation, nor were they exclusively the result of a carefully planned corporate strategy. They slowly evolved out of the ground of scientific and technological competence that existed around the 1920s, during which their development was propelled by, in the first place, a drive to fully uncover the affordances of the electromagnetic spectrum; second, by supervening social necessities such as enhancing the response time of emergency services; and third, by a continuously present demand from potential customers who could afford the technology. In all of these motives, myths of ideal communication played their powerful and strategic discursive role, doing so through justifying the need for investments and large-scale experiments, through orienting developers towards realising popular imagery of what the 'original dream' of mobile wireless communication could be, and through advertisements that upheld the notion that the way that ordinary people experienced mediated communication still needed improvement. We should be careful not to give these myths all the credit, though; regulatory decisions, political motives, and plain technological impossibilities also influenced the process, making it far from self-evident that the mobile communication condition would naturally evolve into the one we now live in.

Still, today the general attitude towards mobile communication devices seems to be, despite heterogeneous reception, one of almost unconditional acceptance, as if the devices truly meet an inherent communicative need. Not only are they integrated into all kinds of everyday activities in enormously large numbers, but their accompanying discourses make it very seductive to think that they indeed present the most logical and natural solution to our communication problems. Many people experience mobile communication devices as primary and indispensable tools for increasing opportunities to connect, socialise, find relief, and gather knowledge, giving credence to the idea that they are perceived as being best suited to fulfil long-held desires for improved communication. What is more, the devices' propensity to also further stress the dark sides of a realised communication utopia is yet another sign that they are part of an evolutionary path towards what we think is communication utopia: relentless connectivity, a blurring of the private and public, terrifying closeness, and anxiety over transparency and accountability are all outcomes of the universalising tendencies that are inherently present in utopian projects. Moreover, by the looks of things, the future of wireless communication technologies will only intensify the communication paradox: when everyone and everything becomes part of the all-encompassing Internet and Internet of Things, we will have to learn to live with living in the 'absolute present' (Allon 2004), in which communication, human relations, and identities will have taken on a radically different meaning.

This is no small matter. It has to be recognised that, within the current state of affairs in the communication media landscape, mobile communication devices have been endowed with the full weight of hopes and desires that have given shape to many previous attempts to come closer to communication utopia. In this sense, there is nothing new about their hyped production, marketing, and reception: we have seen it all before. Nonetheless, we should not be naive to think it is simply business as usual, because mobile communication devices do have an unprecedented and very specific impact on our lives. If basic prerequisites for the constitution of communication utopia are indeed to be understood as being omnipresent, omniscient, and therefore 'omniconnected', then the affordances of mobile communication devices come a long way in meeting those prerequisites, as they virtually guarantee real-time contact with anyone, anything, anytime, anywhere. The paradigm of ubiquitous connectivity has been in the cards for quite some time, and the growing presence and discursive importance of the imagery of mobile communication devices emphatically reflect that. However, omniconnectedness does not necessarily mean that we will, by definition, move towards understanding each other better; with ever-growing opportunities to connect and retrieve and store information, successful communication in the mobile communication condition has de-emphasised the pragmatic dimension of getting to know the other, and has increasingly taken on more instrumental connotations such as establishing and maintaining contact, or collecting data. In this sense, communication's 'good trick' might turn out far more relentless in its consequences than we are able to acknowledge, or are actually willing to see.

Let me conclude this book by saying that the process of writing it has also revealed itself to be an exercise in duly accepting that while the desire to know ideal communication exists, it will never be fulfilled. There have been many times when I just knew that there had to be a better way of expressing my thoughts, that the words were *there*, somewhere, yet not readily available for me to write them down. In those moments of despair, I felt like I was one of the inquisitors in Jorge Luis Borges's 'Library of Babel' (1998): men who, forever unsuccessfully, search the infinite collection of books that hold all knowledge for that single volume that will reveal the fundamental mysteries of humankind. At this last page of my own addition to the library, I know, more than I ever did before I started writing, that all we do is perpetually live through the same tragic yet highly meaningful struggles with finding pure communication.

Notes

Introduction

1. At the time of writing, the discussed video as well as other NTT DoCoMo 'Vision 2010' videos are available for viewing on Youtube (see Seadaniel 2006; Thiagobassolli 2007). From 2008 onwards, as the year 2010 approached quickly and most predictions increasingly looked set to be proved incorrect, subsequent NTT DoCoMo videos have been conveniently renamed Vision 201X and Vision 2020.
2. This discursive strategy employed by NTT DoCoMo very much reverberates that of a memorable AT&T advertisement campaign of the early 1990s, entitled 'You Will'. In these ads, the presentation of several futuristic communication scenarios is accompanied by a male voice-over that asks 'Have you ever...', and at the end of the ads those questions are promisingly and reassuringly answered with 'You will'. See also Rodowick (1995) for a critical analysis of AT&T's future-making rhetoric.
3. This shared use of mobile telephones is mainly visible in developing countries. A poignant example is the 'Village Phone' project, in which micro-entrepreneurs set up shop in rural villages and provide phone calls to the local community on a per-call basis (Grameen Foundation 2011).
4. The hypothesis of indiscrimination is discursively appealing but does not hold empirically; as sociologist Leopoldina Fortunati showed, although the mobile telephone 'expresses a strong capacity for social inclusion' because of 'its disposition as an individual and not domestic or family technology', there are still significant differences to be found in the penetration and possession of mobile telephones between men and women, young people, adults, and elderly, and between countries (Fortunati 2003: 239-241).
5. The predictive power of industry leaders is not to be underestimated. In their seminal article 'Foresight and hindsight: The case of the telephone', Ithiel de Sola Pool et al. (1977) provide an analysis of several reasons why developers are better at forecasting than other commentators, among which the self-fulfilling prophecy plays an important role. Developers are often successful as forecasters, Pool et al. hold, because they 'had the inventions, a vision of how the inventions could be used, and they controlled the businesses that implemented those visions' (Pool et al. 1977: 129).
6. See Paragas (2003) for a critical discussion and a reassessment of the supposed coordinating agency of mobile telephones in instigating and directing the event.
7. Just as they can bring people together in harmony, mobile telephones can also be used to heighten tensions; according to spokesman Bernard Barrett of the International Committee of the Red Cross, for instance, outbursts of violence in Kenya in 2008 were partly fuelled by rumours that were 'spread by cell phone text messages' (CNN 2008).

8. As culture critic Erik Davis writes, 'the magical idea that engineering will create [a more peaceful, virtuous, and wondrous] world is an ominous and tricky dream, though it seems a mighty difficult dream to shake' (Davis 1988: 22).
9. In the context of the citation, Mercury, in keeping with the Greek nomenclature, should of course have been named Hermes. Zeus is known in Latin as Jupiter or Jove.
10. For an elaborate account of the problems that arise when media are 'ontologised', see Boomen (2012).
11. Foucault, in his attempt to cast ideas of historical continuity aside and come to a new form of historiography in *L'Archéologie du savoir* (translated as *The archaeology of knowledge* in 1972), writes that in order to accomplish this we must rid ourselves of notions such as 'tradition': '[I]t is intended to give a special temporal status to a group of phenomena that are both successive and identical (or at least similar); it makes it possible to rethink the dispersion of history in the form of the same; it allows a reduction of the difference proper to every beginning, in order to pursue without discontinuity the endless search for the origin; tradition enables us to isolate the new against a background of permanence, and to transfer its merit to originality, to genius, to the decisions proper to individuals' (Foucault 1972: 23).

1. Discourses of progress and utopia

1. For a broad overview, see for instance the bibliography of secondary sources on the website of the New York Public Library's exhibition on utopian thought, which lists 222 bibliographies, anthologies, conference proceedings, general studies, chronological studies, and thematic studies, most of them written in the last three decades of the twentieth century (NYPL 2000).
2. 'Expectation, hope, intention towards not-yet-become possibility: that is not only a basic feature of the human consciousness, but, when concretely corrected and understood, a basic destination within objective reality as a whole.'
3. One of the more notorious reductionist explanations of human behaviour is found in *A natural history of rape: Biological bases of sexual coercion* (2000), in which Randy Thornhill and Craig Palmer claim that the inclination to rape is the result of evolutionary adaptation. They maintained that their work was not written to excuse the acts of rapists but to better understand how to suppress evolutionary 'left-overs'. Still, the book caused both social and academic outrage (Pozner 2000).
4. The image of human organisms as gradually evolved and immensely complex survival machines of genes was coined by Richard Dawkins in *The selfish gene* (1976). Dawkins himself, following Dennett, also joined a growing array of writers (see Atran 2002; Wolpert 2006) in search of biological explanations of the need to believe by writing *The God delusion* (2006), thus emphasising the present-day persistence of understanding systems of hope and belief in evolutionary terms.
5. An excellent account of the 'art and critique of forgetting' is found in Harald Weinrich's book *Lethe* (2004), in which he shows that forgetting – even though we cannot put it into an imperative – can have a strong social and psychological significance by providing the enticing image of 'an opportunity to start all over from the beginning' (Weinrich 2004: 165).

6. Campbell has been criticised, most notably by Pearson and Pope (1981), for having an old-fashioned view on gender relations, placing women in traditional roles of mothers and wives, and men as heroes.
7. Religions form the domain par excellence where the sublime, myths, purpose, hopes, and beliefs find their expressions. According to Paul Wong, Victor Frankl held that '[r]eligion encompasses the ultimate meaning, super-meaning, as well as God' (Wong 2001). Also see the work of Claude Lévi-Strauss on religion and mythology in *Anthropologie structurale* (translated as *Structural anthropology* in 1963) and *Paroles donnés* (translated as *Anthropology and myth: Lectures*, 1951-1982 in 1987).
8. Note the similarities here with the ubiquitous and binding power called 'the Force', found in the six movie episodes of STAR WARS (1977-2005). Director and writer George Lucas was greatly influenced by Campbell's work, and admitted that he had rewritten parts of the scripts for the original STAR WARS trilogy after having read *The hero with a thousand faces* (Larsen & Larsen 2002; Murray 1997: 186).
9. For a detailed breakdown of followers of the world's religions, see Wikipedia's entry on 'Religions by country' (Wikipedia 2011).
10. Political theorist Vincent Geoghegan notes: 'The religious traditions of humanity contain the gold-bearing seams of utopian hope – in many ways the most distant from existing reality, but also grounded in the most intimate of human concerns' (Geoghegan 2007: 112).
11. Bloch wrote *The principle of hope* during his exile in the U.S. after (and because) he had to flee Nazi Germany. His work can be understood as a reaction against the fact that the Nazi regime, in his view, misused utopian rhetoric – the use of the term 'Third Reich' being the most obvious example – and institutionalised it in a restrictive way. See also Kellner (1997a).
12. The argument he used to rationalise the arrangement of an unconscious that is supplemented by a not-yet-conscious was that '[w]hat was never conscious cannot become unconscious' (Bloch 1919: 355 cited in Bloch 1988: xxxi).
13. J.B. Bury, in his landmark study *Idea of progress* (1920), maintained the view that a sense of progress is only possible when one is convinced of history moving at a certain pace and indefinitely into the future, and that this is typical of the modern frame of mind. Historian Sidney Pollard writes that 'the sense of history itself, including the idea of human progress, was absent in classical times, and could grow only after the mental fetters inherited from them had, at least in part, been broken' (Pollard 1968: 1).
14. The word 'progress' in classical antiquity was predominantly associated with 'pushing forward' (from the Greek προκοπή [prokope]) or 'an advance' (from the Latin *progressus*); both movements that were either cyclical or eventually finite (Wiener 1973: 624).
15. Hesiod himself actually wrote of 'races' instead of 'ages', and described five stages instead of four, including an uncharacteristically non-metal Heroic Age in-between the Bronze and Iron Ages. Later accounts reorganised the scheme by changing the nomenclature and dropping the Heroic Age (see Baldry 1952; 1956).
16. According to the classification developed by Charles van Doren in *The idea of progress*, Hesiod belongs to the 'regress authors', most probably because Hesiod wrote he would rather have died before, or had lived after, the Age of Iron instead of during his own time period (Van Doren 1967: 9).

17. Ernst Bloch as well was greatly influenced and fascinated by Joachim de Fiore's apocalyptic spirituality. Also see Geoghegan (1996: 90-103).
18. The place literally does not exist, its major city is called Amaurot, which means 'phantom', and the protagonist is named Raphael Hythloday, which means 'speaker of nonsense' in Greek.
19. For a comparison between Plato and More, see chapter six of *After Augustine: The meditative reader and the text* (Stock 2001).
20. For elaborate studies on the history of utopian settlements in Northern America, see *Heavens on earth: Utopian communities in America 1660-1880* (Holloway 1966) and *A visionary nation: Four centuries of American dreams and what lies ahead* (Karabell 2001).
21. Works that aimed to answer questions of why there is a world and why it is the way it is were most notably written by Gottfried Wilhelm Leibniz. According to historian Ayval Ramati, Leibniz, in his 'De Rerum Originatione Radicali' ['On the ultimate origination of things'], asserted that the world 'had not reached its perfection on the day of creation', and that progress should be 'understood [...] as continuous eternal change toward perfection' (Ramati 1996: 433).
22. See the introduction to Smith and Marx (1994: xi) for the argument that the 'deterministic view of technology is a pervasive theme of the mass media nowadays'.
23. Glenn Negley, in his bibliography of utopian literature, lists 239 utopian works written from 1650 to 1800 (Negley 1977: 224-225); Bronislaw Baczko counts around 80 '*voyages imaginaires*' in the domain of French language from 1676 to 1789, but while referring to even more impressive estimates, Baczko maintains that 'on average more than 10 and a maximum of 30 "*voyages imaginaires*" in certain years' could have been published (Baczko 1978: 47, author's translation). Even if many of these texts were of inferior quality, and even if imaginary voyages do not necessarily describe utopias, as Baczko cautions, the total amount still adds up to much more than ever before (ibid.: 47-48).
24. Hegel's historiography has been criticised for being Eurocentric, racist, and totalitarian (Dussel 1993).
25. Desroche notes the intertwined relationship of utopian thought and hope: 'Are not utopia and hope, in particular, twin sisters? In utopia there is the hope of a different society. In hope there is the utopia of a different world. In both of them there is the strategy of alterity' (Desroche 1979: 23).
26. An elaborate argument that recurrent apocalyptic thinking is of all times is found in Richard Abanes' *End-time visions: The road to Armageddon?* In this book, Abanes, himself once a cult member of The Way International, extensively shows how '[h]umanity's preoccupation with doomsday may go back as far as our earliest evidence of communal living' and that '[s]everal end-time scenarios from centuries past sound strikingly similar to each other' (Abanes 1998: 159).
27. The incessant drive towards the unreachable goal is the core of the paradox. As Jean Baudrillard, in his depiction of America as the hyperreal achievement of utopia, remarked: '*A utopia achieved is of course a paradox*, as utopias are not meant to be achieved' (Baudrillard 1993: 245, emphasis in original).
28. See also Douglas Hofstadter's classic *Gödel, Escher, Bach: An eternal golden braid* (1979), in which he describes the existence in life of *strange loops* (instances of self-reference that

lead to paradoxical situations), and shows that these loops can not only be found in math, perspective drawings and music, but also – and this is his main argument – in the very essence of conscious existence itself.

29. All following citations from this book are the author's translations.
30. Crombag and Van Dun call this fourth solution 'rechtvaardigheid', which literally translates to justice, but here refers more to 'the solution that makes sure everyone has access to and control over what is his or hers, and cannot have access to or control over what belongs to someone else without permission' (Crombag & Van Dun 1997: 28).
31. The difference between a dystopia (literally meaning 'bad place') and an anti-utopia is, as Wegner explains: '[T]he former presents a critique of the limitations of a specific form of imagining place, the latter a rejection of this cognitive act altogether' (Wegner 2002: 152-153). I will refer to both narrative forms in the context of this subsection, as they equally confront the utopian text with its inherent conflicts.
32. Karl Mannheim writes in *Ideology and utopia*: 'The representations of a given order will label as utopian all conceptions of existence which *from their point of view* can in principle never be realized' (Mannheim 1936: 196, emphasis in original). Although this observation leaves open the conclusion that 'to think utopian' has, from the dominant perspective, always implied having a naive or unrealistic worldview, Mannheim adds a historical component: 'According to this usage, the *contemporary* connotation of the term "utopian" is predominantly that of an idea which is in principle unrealizable' (ibid., emphasis added). Frank and Fritzie Manuel, assessing the body of critical studies of utopias in the twentieth century in their book *Utopian thought in the Western world*, remark that 'to many observers [utopianism] was a corpse', a '*Schimpfwort*' for 'outdated and historically superseded fantasies' (Manuel & Manuel 1979: 10-11).
33. This is not to say that ideal societies are always depicted as artificial constructions and as distinct from living conditions found in nature's wilderness; ascetic utopias for instance seek measures to bring about a return to a Rousseauian natural paradise.
34. Wegner as well observes this necessity in his analysis of Orwell's text: '*Nineteen Eighty Four* [...] inexorably drags the reader to the conclusion that every effort to effect a total change of the present, to institute a utopia [...] invariably gives rise to total systems of domination, systems wherein, ultimately, even the potential for change might be eliminated' (Wegner 2002: 191).
35. The protagonist of the novel, Winston Smith, works for the Outer Party in the Ministry of Truth, where he rewrites historical documentation so that it always fits the Party's policy.
36. Following Bloch in recognising the productive quality of dialectical thought, Herbert Marcuse, in his work on the 'Orwellification' of the language of twentieth-century political discourse in *One-dimensional man* (1964), advocates a constant refusal as the proper starting point for political activism against totalitarian oppression. Such a refusal is quite clearly visible in the narrative of *We*, but also in popular reception of *Nineteen eighty-four*. As Douglas Kellner writes in his essay on Orwell and Marcuse, 'the very antithesis between "optimism" and "pessimism" as opposing political mind-sets is a specious one and supposedly pessimistic ideas can be productively used to mobilize people against oppression' (Kellner 1997b).

2. Communication ideals, communication woes

1. Socrates saw his philosophical work as a form of intellectual midwifery, or *maieutics*, giving birth to ideas and insights that promote critical self-reflection.
2. Most notably in works such as *Apology*, *Cratylus*, and *Phaedrus*.
3. The Socratic dialogue does present us with questions concerning equality between teacher and student; are they indeed equal because the teacher has to admit that she does not know the answers to the questions she poses, and therefore does not know more than the student? Or, as Richard Elmore (2005: 279) notes, is the teacher being manipulative, because she 'pretends not to know what she knows in order to engage the student in a line of questioning that leads the student to a predetermined destination described by the teacher's knowledge'?
4. Jürgen Habermas' work on communicative action, arguably one of the twentieth century's most influential endeavours to theorise the ideal speech situation, has also been identified as having a strong bias in favour of face-to-face dialogue, instead of mediated conversation (see Page 1996; Schudson 1997).
5. Where 'Basic' is an acronym for 'British American Scientific International Commercial'.
6. Ogden gives some examples of how Basic English works to transpose words and sentences into their elementary parts: '*Disembark*, for example, is broken up into *get off a ship*. I *am able* takes the place of I *can*; *shape* is covered by the more general word *form*; and *difficult* by the use of *hard*' (Ogden 1935: 14).
7. Especially among the philosophers who attended meetings of *Der Wiener Kreis* [*The Vienna Circle*] in the 1910s and 1920s, with participants such as Moritz Schlick, Otto Neurath, and Hans Hahn, the use of logical analysis and symbolic logic (based on the mathematical work by Bertrand Russell and Ludwig Wittgenstein) was aimed at leaving the ambiguity of natural language behind and arriving at unambiguous solutions to philosophical problems.
8. Shannon's proof for the possibility of error-free data transmission was extremely useful for Bell's research division Bell Telephone Laboratories, where Shannon worked on improving the efficiency of transmitting information in telephone switching circuits.
9. The notion that communication technologies are not about transferring meaning but about perfecting the fusing of data streams into a single 'endless loop' of 'absolute knowledge' can most notably be found in Friedrich Kittler's (1999: 2) work.
10. The basis of Lasswell's model of communications comes down to the question of 'who says what, through which channel, to whom, with what effect?' (Lasswell 1948: 37).
11. The word 'informationalisation' is translated from the title of chapter 7 in *Cyberspace odyssee* [*Cyberspace odyssey*] (2002). The original word is 'informatisering', which is actually translated as 'computerisation' in the leading Dutch *Van Dale* dictionary. This shows how much the term 'information' has become entangled with a computational process. See also *My mother was a computer* (2005) by N. Katherine Hayles for a discussion of how the 'regime of computation' has come to determine our present-day worldview (Hayles 2005: 15-38).

12. Captured in the precept 'Therefore all things whatsoever ye would that men should do to you, do ye even so to them: for this is the law and the prophets' (Matthew 7: 12)
13. 'And when much people were gathered together, and were come to him out of every city, he spake by a parable: a sower went out to sow his seed: and as he sowed, some fell by the way side; and it was trodden down, and the fowls of the air devoured it. And some fell upon a rock; and as soon as it was sprung up, it withered away, because it lacked moisture. And some fell among thorns; and the thorns sprang up with it, and choked it. And other fell on good ground, and sprang up, and bare fruit an hundredfold. And when he had said these things, he cried, He that hath ears to hear, let him hear' (Luke 8:4-8).
14. Most of the Frankfurt scholars' condemnations of the dark sides of mass media were first informed by their encounters with state-controlled propaganda in Nazi Germany, and then, to their dismay, by their observation that in democratic countries such as the U.S. mass media were similarly used to manipulate minds.
15. In the 1970s and 1980s, under the influence of scholars such as Jay Blumler and Elihu Katz (1974) and Stuart Hall (1980), critical analyses turned more to questions of what people did with media, than of what media did with people. This shift towards seeing audiences as 'active' has, since the rise of the Internet in the 1990s, been followed by a movement towards identifying them as 'participatory' (Jenkins 1992; 2006). For a critique of the highly optimistic connotations of the participatory position, see Schäfer (2011).
16. See for instance Nealon & Irr (2002) for a contemporary 'rethinking' and reinstatement of the critical theory of the Frankfurt School.
17. From the Greek νους, which means 'mind'.
18. The exact influence of Teilhard de Chardin on McLuhan's work is, despite clear similarities, difficult to gauge. According to Tom Wolfe, the journalist who popularised McLuhan's work in the 1960s and who became a close friend, McLuhan 'acknowledged his tremendous debt to Teilhard de Chardin' privately, but never publicly (Wolfe 2004). This is disputed by Donald Theall, who says that 'McLuhan quotes or mentions Teilhard on a number of occasions in his *Gutenberg Galaxy*; and among those remarks he clearly indicates the reservations he has concerning Teilhard's writing' (Theall 2006: 697).
19. Also see Margaret Wertheim's *The pearly gates of cyberspace*, in which she investigates the captivating yet problematic techno-religious dreams that portray cyberspace as 'a utopian arena of equality, friendship and virtue', culminating in 'a technological substitute for the Christian space of heaven' (Wertheim 1999: 16).
20. On his website Bloom states in 'The omnologist manifesto' that 'Omnology is a science, but one dedicated to the biggest picture conceivable by the minds of its practitioners. Omnology will use every conceptual tool available – and some not yet invented but inventible – to leapfrog over disciplinary barriers, stitching together the patchwork quilt of science and all the rest that humans can yet know' (Bloom 2001).
21. In evolutionary biology, the theory of group selection has long been regarded as less suitable for explaining the behaviour and evolution of organisms; instead, individual selection, which is based on a gene-centred view, has been favoured. In the last dec-

ade or two this has changed, and group selection (or multi-level selection as it is now called) has gained favour (see Sober & Wilson 1998; Boehm 1999).

22. See Russell (1995) and De Kerckhove (1997) for examples of how the notion of the emergence of a global brain has been informed by the new communication and information processing technologies of the late twentieth century.
23. Borgmann's fear of a loss of materiality in information is conspicuously evident from the ominous title of his book, *Holding on to reality* (1999).
24. It is no coincidence that he mimics here the eccentric cyberpunk style of William Gibson's *Neuromancer* (1984); Lévy's *Collective intelligence* fits the genre of books that portray a futuristic digital universe and use vivid language to sweep readers off their feet.
25. A particularly poignant display in popular culture of the inevitable loss of identity in universalising processes is found in the science fiction TV and movies series STAR TREK, in which a technologically enhanced race called the Borg assimilates individuals into its collective hive-mind in a quest for ongoing improvement. Communication in the hive-mind is instant; every Borg entity knows exactly what all the other Borg entities in the Collective know. For a discussion on the role of the Borg as the representation of the dystopian future of humankind, especially in relation to communication, see Kavanagh, Keohane & Kuhling (2001: 14-18).
26. In explaining the fundamental anxiety of the universalising aspects of utopia, Fredric Jameson writes that it is 'the fear of losing that familiar world in which all our vices and virtues are rooted (very much including the very longing for Utopia itself) in exchange for a world in which all these things and experiences – positive as well as negative – will have been obliterated' (Jameson 2005: 97).
27. At the 2003 ARS Electronica Festival in Linz, Lévy presented a bewildered and bemused audience with an elaborate system consisting of 96 different behavioural patterns with which he intended to capture the essential meaning of all symbolic activities taking place on the Internet.
28. See also the conclusion to Eco's *The search for the perfect language* (1995: 337-353), in which he notes that myths of a universal understanding are inherently countered by everyday experiences of quirks in any language, and that it is precisely because of differences in languages that interesting projects (such as the long-term project of building a united Europe) are initiated.
29. Elimination might not be the correct word as it implies that technical noise can always be entirely taken out of the equation, but, following Shannon, even if there is technical noise in a channel it can be overcome to such an extent that the original message is received unchanged, as if there were no noise at all.
30. It is commonly assumed that Kierkegaard was anti-Hegelian, but according to Jon Stewart his criticism was more directed to contemporary Danish Hegelians than to Hegel himself. Kierkegaard's criticism was that advocating abstract universal thought tends to over-marginalise the difficulty of individual existence (Stewart 2003).
31. There is no single existentialist doctrine or school of thought; many philosophers who are traditionally categorised as existentialists – Heidegger, Kierkegaard, Nietzsche, Sartre, Jaspers – had widely diverse backgrounds, beliefs and opinions. They do, however, all emphasise that people are thrown into the world, and thus have to experience

their existence as an interdependence between themselves and others (Reinhardt 1952).

32. Steven Shaviro makes a similar observation in his analysis of how proponents of artificial intelligence programmes such as Vernor Vinge and Ray Kurzweil propagate the coming of what they call the 'Singularity', a rupture in human history that is brought about by ever-accelerating technological change. In this mode of reasoning, Shaviro notes, artificial intelligence is presented as 'the next step in evolution [...] and our only viable option is to Get With The Program' (Shaviro 2003: 121). Here we see the linear argument at work again, framing the Singularity as the inevitable final goal of 'a necessary and continual upward progression to a state of transcendence' (ibid.: 122).

3. The rise... and rise of media technology

1. The rise of the industrial age did not exclusively engender ideas of progress, of course; pollution, lost jobs, hard labour, poor housing conditions, and an increase in stressful environments led to pessimistic expectations as well. Still, new technologies commonly perpetuated myths of progress.
2. Vice versa, having a technological determinist disposition does not always mean seeing things in positive progressive terms; see the work of Jacques Ellul (1964; 1980; 1990) for a predominantly pessimistic determinist view on technology.
3. The academic consensus in the past few decades has been that 'the term "deterministic" tends to be a negative one', and some even use it as 'a term of abuse' (Chandler 2002).
4. See for instance Grint & Woolgar (1997) for the argument that technological artefacts can be 'read' in virtually unlimited ways.
5. In the words of historian Michael Smith as he questions the persistence of technological determinism, it is necessary to come to understand '[h]ow something so demonstrably wrong-headed [can] continue to sway adherents' (Smith 1994: 39).
6. As Chris Harman has noted on Marx's claim: 'The summation is crude. It is also historically inaccurate. What accompanied the rise of European feudalism after the 10th century was not the spread of the handmill, but its replacement over the centuries, the watermill – and the watermill then went on to play an important role in the genesis of industrial capitalism. But Marx's central point was correct' (Harman 2004).
7. Unlike Hegel, however, Marx thought that the activity of *Geist* ought to be understood in more pragmatic terms; its self-realisation was to be found in 'the actual social and political institutions that encompass man', instead of in perfected abstract knowledge (Bernstein 1971: 40).
8. See for instance William H. Shaw (1979: 172-175), who is nonplussed by the fact that many Marxists tend to condemn as vulgar the view that Marx's theory of history is one of technological determinism, while such an interpretation of Marx, Shaw argues, 'does not impute to him a wildly implausible theory', and can be 'positively helpful' in 'stimulating fresh and useful research, uncovering new facts, and promoting empirical hypotheses and models.'
9. Also see MacKenzie (1984) for an extended argument why Marx's writings ought not necessarily be seen as technologically deterministic.

10. The multiscalar approach suggests following and describing interactions between micro-scale, meso-scale, and macro-scale actors. The meso scale here should be understood as 'the region conceptually intermediate between the macro and the micro', in which 'institutions intermediate between the firm and the market or between the individual and the state' reside (Misa 1994: 139). I primarily want to stress here that I think it is productive to go back and forth between the various levels, and not, as might be suggested when looking for a balanced view between macro and micro levels, to focus solely on the meso level.
11. As an example of a technological artefact that is presumed to be universally adopted and recognised as the quintessential symbol of progress when it was invented, Basalla refers to the wheel, which, in fact, is 'a culture-bound invention whose meaning and impact have been exaggerated in the West' and 'is not a unique mechanical contrivance necessary, or useful, to all people at all times' (Basalla 1988: 11). See also Wright (2000: 46).
12. For a comprehensive account of the dark sides of technology see Edward Tenner's *Why things bite back: Technology and the revenge of unintended consequences* (1996), in which he lists 'revenge effects' of, amongst other examples, the computerisation of the office, which has led to health problems such as carpal tunnel syndrome.
13. The popularity of evolutionary theories in attempts to explain a myriad of behaviour phenomena has led to many uneasy and downright inadequate applications of evolutionary principles, so it is important here to stress the possibilities *and* the limitations (see Laland & Brown 2002).
14. The list can hold many more items; for instance, historian Alex Roland finds that demands are imposed by external factors (such as perceived need, resources, national and regional style, ideology and philosophy, politics, military economics, and science) and internal factors (such as entrepreneurs and translators, inertia and momentum, systems, institutions, plateaus, standardisation, and secrecy) (Roland 1992: 84-90).
15. See Whyte (2007: 46) for a discussion of the difficulties of using evolutionary theories – in which intentionality plays a marginal role at best – to explain 'the deliberate and intentional nature of design practice'.
16. Thomas P. Hughes, in his work on the electrification of Western society, sees inventions and technological developments mainly as the result of attempts to correct 'reverse salients', a term with which he describes that, when technology is viewed as a goal-seeking system, some components of that system are perceived 'to fall behind or out of line' and thus are redefined as 'critical problems' that need to be solved (Hughes 1983: 80). He adds: 'Outstanding inventors, engineers, and entrepreneurs usually have a record of defining and solving such problems' (ibid.).
17. In historiography, there seems to be a repetitive tendency to construct theoretical concepts with which to understand and explain differences and similarities in constellations of cultural and technological artefacts. Memes and topoi are most suitable for my argument here, but one can also think of tropes, paradigms, clichés, and so forth (Vries 2006).
18. The contention that imagined media can still be interesting research objects is subscribed to by Huhtamo as well, when he notes that 'unrealized "dream machines", or

discursive inventions (inventions that exist only as discourses), can be just as revealing as realized artefacts' (Huhtamo 1994).

19. The assumption that 'true' identities of media exist, is, for instance, poignantly visible in Paul Levinson's 'anthropotropic' view on media evolution, with which he explains how 'the balance which we orchestrate in our media is but a reflection of our own internal, pre-wired balance', a balance that '[humans] come programmed with' and consists of 'time and space projectors that we shine upon the world' (Levinson 1997: 60).
20. Kessler derives his understanding of the dispositif mostly from the work of French philosophers Jean-Louis Baudry and Michel Foucault. Often (but problematically) translated into English as 'apparatus', the dispositif in Baudry's sense implies a certain viewing situation of spectators, for instance of those watching a movie in the cinema (Kessler 2006).
21. In one particular and anecdotal story about Morse, the origin of his drive to improve communication is traced back to 1825, when he was on a business trip and only learned of his young wife's sudden death a week after it had happened and she had already been buried. After that, so the story goes, '[h]e never wanted anyone to go through the pain that he had endured and so set about perfecting an easy to use message system' (Luscombe 2007).
22. For a comparison between the telegraph and subsequent communication technologies, see Standage (1998: 205ff). Standage names the telephone, the fax machine, and especially the Internet as technologies that have much in common with telegraphy networks.
23. From the 1850s to the 1920s, Britain dominated the world's cable telegraphy industry, connecting almost all of its colonies into a single communications network. As a 'quintessential technology of empire', telegraphy helped shape and maintain Britain's control over vast territories (Hunt 1997: 313).
24. Unitel Telegram Services, a Swiss-based private company, now owns, handles, and/or operates many former national telegram services after they were officially abandoned by state companies and large commercial enterprises at the beginning of the twenty-first century. In the Netherlands, it costs €27 to send a telegram and have it delivered in person (Telegram Service 2008).
25. A string telephone is also known as a 'lovers' telephone': by connecting two paper cups with a tight string, sound waves can travel along that string.
26. Only in 2002 was it officially resolved by the U.S. House of Representatives that 'the life and achievements of Antonio Meucci should be recognized, and his work on the invention of the telephone should be acknowledged' (U.S. House of Representatives 2002: 2).
27. Reis' telephones functioned on the principle of 'making and breaking' an electrical current, in fact mimicking how telegraphy works. For speech to be converted into electrical signals, however, there has to be a continuous, variable contact between the transmitter and the electrical circuit. It is believed that Reis' experiments would eventually have led to speech-transmitting telephones, had he not died prematurely in 1874 (Thompson 1883: 8).

28. Popular myth has it that Gray's documents were delivered at the U.S. Patent Office at a later time than those of Bell. According to Evenson (2000: 68-69), however, Gray's caveat was handed in *before* Bell's patent application, but the latter was given priority. Bell's dossier was thus processed earlier, relegating Gray to the sidelines of the telephone industry.
29. Claude Fischer notes that advertisements for telephone services often stressed increased security and practicality, and promised that the telephone would bring people closer together in a better understanding of each other (Fischer 1992: 163-164). Such optimistic notions of improvements in communication are of course ideologically biased; as Carolyn Marvin writes, the idea that '[i]nstantaneous electric communication augured a universal language [...] and global harmony' was a 'distinctly Anglophile solution' (Marvin 1988: 193).
30. In 'Misunderstanding and its remedies: Telephone miscommunication' (1991), Kent Drummond and Robert Hopper note that a telephone conversation is of course constrained to sounds, making the telephone a 'leaner' medium than face-to-face speaking. However, they also note that there are more similarities than differences between the two, and that both of their conversational strategies to circumvent or repair misunderstandings are quite alike (Drummond & Hopper 1991: 314).
31. It is of course still possible to use the telephone to call in and obtain information, for instance in the form of time telling or weather forecast services, but such information is often automatically 'narrowcasted' to the caller instead of being broadcasted continuously.
32. The rate of growth of the telephone networks was not uniform across countries. Technical and organisational issues often caused delays or even cancellations of both national and international network expansions. In addition, in countries where telephone networks were mostly or completely owned and controlled by the state, political interference not seldom hampered the speed of development. For an elaborate historical account of the construction and maintenance of the Dutch telephone network, see Onno de Wit (1998); for the French perspective, see Catherine Bertho-Lavenir (1988); for the German telephone history see Frank Thomas (1988).
33. Carolyn Marvin would characterise Casson's hopeful expectation as a typical expression of a strong belief in new media's promise of 'complete cross-cultural understanding', which would break down barriers and bring 'appreciation and friendliness' to all cultures (Marvin 1988: 194).
34. As physicist John Ambrose Fleming explained: 'The physical effect we call light, and that which we have up to the present moment merely called electric radiation, are really identical in nature, and both consist in waves propagated through the space-filling æther, the only difference between them is in wave-length and wave-amplitude' (Fleming 1902: 254).
35. The existence of the æther was first put into question by experiments of Albert Michelson and Edward Morley, and subsequently rejected in 1905 by Albert Einstein's special theory of relativity (Peters 1999: 103).
36. The developments in this constellation were in no sense unilinear, as Flichy notes: 'What appears today as a series of naturally articulated steps is, in reality, the history of a difficult passage from one domain to another; from science to technology (and

vice versa), from the military to telecommunications, from commercial information to entertainment, and so forth' (Flichy 1995: 100).

37. Crookes was an English chemist and amateur physicist who was very much influenced by Spiritualist views on communication.
38. See Blaine (1905) for a contemporary account of the development of wireless telegraphic devices.
39. The headstrong hero myth is quite prominent with Marconi. According to Orrin E. Dunlap, 'Marconi never gave up. Skeptics could not dent the armor of his faith. Always he plodded ahead under the slogan, 'Occorre progredire!' – 'We must progress!' He was endowed with a plastic power; he mastered circumstances' (Dunlap 1937: 34).
40. 'DX' stands for 'distance'. 'Dxing' meant trying to make contact with other radio operators who were as far away as possible (Douglas 1987: 307).
41. Jenkins and Baird were of course not alone at the time in arriving at functional television systems; other names include Max Dieckmann in Germany, Frank Gray and Philo Farnsworth in the United States, Eduoard Belin in France, and Kenjiro Takayanagi in Japan (Abramson 1987: 73-107). The experiments of Jenkins and Baird, however, remain the most prominently known.
42. It has to be noted that in these movies two-way television is not always portrayed in a positive light; in METROPOLIS and MODERN TIMES, for instance, the device is used by an ominous ruler/boss to keep an eye on his workers and tell them when they have to speed up their activities.
43. For an excellent account of failure in technology in general and of the failure of AT&T's Picturephone in particular, see Kenneth Lipartito (2003). Lipartito as well stresses that studying dead media can be as revealing as looking at successes: 'Failed technologies, far from being dead ends or even mere cautionary tales, may persist well beyond their material life. They may reinforce rather than undermine technological paths, even when those paths are questionable or undesirable. Failures, the story of Picturephone suggests, can echo like footfalls down corridors not taken, leading us to the present' (Lipartito 2003: 53).
44. See Sixma (2007) for a detailed overview of the many experiments with videophones up until today, as well as of the strategic uses of the topoi of mediated closeness and immersive presence that have always accompanied the development of videophone systems.
45. As Manuel Castells noted: 'While the media have become indeed globally interconnected, and programs and messages circulate in the global network, *we are not living in a global village, but in customized cottages globally produced and locally distributed*' (Castells, 1996: 341, emphasis in original).
46. Together with his friend Henri La Fontaine, Otlet was a prominent advocate for the creation of the League of Nations and its educational body The International Commission on Intellectual Cooperation (which later became UNESCO).
47. All translations from Otlet's originally French work are the author's.
48. Otlet imagined all the world's knowledge to be centrally stored on standardly formatted three-by-five-inch cards. Later he tried using microfilm instead. Attempts to create and maintain a central, universal library, one which he called the Palais Mondial

or Mundaneum, were largely unsuccessful due to excessive maintenance needs and an uncooperative government. Also see Rayward (1991).

49. As another example of how the evolutionary nature of technology development can orient independent experimenters and thinkers towards the same 'good tricks', Bush's design for a microfilm selector was predated by one that was patented and developed by the Russian-born German photography expert Emanuel Goldberg in 1931 (Buckland 1992).
50. According to Internet Systems Consortium (ISC), the number of hosts in October 1985 was 1,961. In October 1990, this number had grown to 313,000; in July 1995, to 6,642,000; in July 2000, to 93,047,785; and in July 2005, to 353,284,187. As of January 2011, a total of 818,374,269 Internet hosts had been counted (ISC 2011).
51. Grassroots hobbyists' computer networks had already existed since the end of the 1970s in the form of Bulletin Board Systems (BBSs), computers that ran software that could dial into other computer systems using regular phone lines.
52. See Negroponte's *Being digital* (1995) for a particularly telling account of how atoms and bits are believed to be merging into a single stream of information.
53. The dot-com bubble at the beginning of the 2000s, itself the product of an overhyping of the assumed limitless possibilities of the digital new economy, is often mentioned as a prime example of what was wrong with how the Web was used. See O'Reilly (2005) for an illustration of this type of reasoning.
54. Because its emphasis on increased possibilities for collaboration, participation, and sharing has over the years become best visible in new software platforms that display and interconnect all kinds of users' profile information, the Web 2.0 metaphor is nowadays quickly being replaced by the term 'social media'.
55. For a well-founded criticism of present-day software's increased reliance on the disclosure of personal information, see 'The externalities of search 2.0: The emerging privacy threats when the drive for the perfect search engine meets Web 2.0' (Zimmer 2008). Also see Lanier (2010) and Schäfer (2008).

4. Mobile communication dreams

1. Compounding 'motor' and 'ola', implying 'sound in motion'.
2. There were some sporadic experiments with systems that could interact with the fixed telephone network; in 1924, Bell Labs tested a car-bound radio telephone that could place two-way voice calls with land-based telephones (Farley 2006a).
3. In the documentary HOW WILLIAM SHATNER CHANGED THE WORLD (Jones 2005), Martin Cooper names the popular 1960s science fiction television series STAR TREK as a major influence on his thinking. Commenting on how he was fascinated by how Captain Kirk's versatile and universal 'communicator' operated, he says: 'that was not a fantasy to us, [...] to me that was an objective: [...] the idea of being able to talk not only to the next floor of the *Enterprise*, but to people on the planet'.
4. 'Auto' can also mean 'car'. The ATF-2 system still made reference to cars because of its relation to the first ATF netwerk, but the Carvox 2453 mobile telephone model could be used outside of a car (Groen 2006).

5. Such as the availability of micro-electronic components, a consumer market that was susceptible to novel and personalised technologies, and companies and business structures that were aimed at capitalising on the opportunities that arose out of that consumer market.
6. Numbers from Eurostat show that in the current 27 EU countries in 1998 there were about 20 mobile telephone subscriptions per 100 inhabitants; this figure rose to 78 in 2003 and to 125 in 2009 (Eurostat 2011).
7. The GSM Association predicted in a 2010 press release that the number of global mobile connections would reach 6 billion 'in the first half of 2012' (GSM Association 2010).
8. Paul Levinson's *Cellphone: The story of the world's most mobile medium and how it transformed everything!* (2004) provides a good example of the argument that mobile communication devices are a perfect fit for 'natural' human communication needs.
9. This section has, in an edited version, previously been published as 'The vanishing points of mobile communication' (Vries 2008).
10. It has become a regular and accepted practice for telecom operators to disguise high towers of mobile base stations as trees in order to conceal them from view, or integrate telecommunications equipment into highly placed infrastructural objects such as bell towers or, in a striking expression of 'angelic communication', even church crosses. See Stealth (2008) for some telling examples.
11. Also see Robison (2003) for the argument that the terminology of mobile communications is guided by 'the *ideology of telepathy*: an ultimate, implied goal of total convergence between machine, man, knowledge and communication' (Robison 2003: 175, emphasis in original).
12. Also see Churchill and Wakeford (2002) for examples of how advertisements for wireless mobile communication devices explicitly state goals of 'unlimited *access to others* and unlimited *access to information*' (Churchill & Wakeford 2002: 163, emphasis in original).
13. The prospect of today's young generations being always connected has prompted some scholars to state that 'the youth, they truly are the Borg' (Ahonen 2006). While this is of course an exaggerated claim, mainly because we can safely assume that most young people are still very much unique individuals, it does show that the Borg are iconic and easily understandable examples of what a radically increased connectedness might entail. Also see Gunkel (2000) in this respect.
14. Italian philosopher Gianni Vattimo described the ideal of building a world community through unlimited communication as one that aimed for a complete 'transparent society', in which communication media would offer 'a kind of concrete realization of Hegel's Absolute Spirit: the perfect self-consciousness of the whole of humanity, the coincidence between what happens, history and human knowledge' (Vattimo 1992: 6). In his work, Vattimo exposes this ideal as illusory, stating instead that in the media society we will become aware of the significance of plurality and the non-existence of 'perfect knowledge' (ibid.: 7).
15. For explorations of the debate on whether mobile communication technologies inherently create a more individualised society, see Arnold (2003: 247ff), Geser (2004) and Horst & Miller (2006: 81ff).

16. Also see Rakow and Navarro, 'Remote mothering and the parallel shift: Women meet the cellular phone' (1993).
17. For a rather striking example of how the metaphor of the leash is readily used to give expression to how parents might think of improved communication as an antidote to their anxiety, see the 'Safe & Sound: a wireless leash' technical proposal by MIT scholars Natalia Marmasse and Chris Schmandt: 'Safe & Sound [...] is a location-based information system that allows parents to monitor their children's location. The child's phone continuously streams location information to the parent's phone. A voice channel between the parent and child enables negotiating with, and perhaps educating of, the child. As the title of the project implies, *we emphasize both security and communication for this solution to a common parental concern*' (Marmasse & Schmandt 2003: 726, emphasis added).
18. See Ling (2004: 119-121) for further illustrations of how issues of parental control are often part of processes of emancipation of adolescents.
19. For similar diagnoses of the risks of losing control over new mobile technologies, see *Surveillance as social sorting: Privacy, risk, and digital discrimination* (Lyon 2003) and 'The wireless leash: Mobile messaging service as a means of control' (Qiu 2007).
20. The idea that mediated communication stresses and problematises the physical absence of the other was of course already articulated by Socrates, as noted in chapter 2. For a detailed account of how telephonic conversations can separate people while being connected (and of how this affected dramatic cinematic structures in the early twentieth century), see Tom Gunning's essay 'Heard over the phone: *The Lonely Villa* and the de Lorde tradition of the terrors of technology' (1991).
21. Similar stories of 'ringing body bags' could be heard after the Madrid terrorist attacks in 2004 (BBC News 2004) and the Virginia Tech shootings in 2007 (King 2007). See also the song 'Cell Phones Ringing (In the Pockets of the Dead)', released by singer-songwriter Willie Nile in 2004 after the Madrid bombings.
22. See Wellman (2002) for the argument that new media technologies increasingly individualise people.
23. The practice of seeing the initial use of new technologies simply in terms of a continuation and improvement of that what worked well in old technologies is what Scott Jenson calls 'legacy vision' (Jenson 2005: 307). Because the development of the Internet and the subsequent uptake of the World Wide Web run roughly along the same time line as the development of cellular telephony and its spectacular rise as the medium of choice to communicate on the move, Jenson says, a 'manic belief' took hold in the late 1990s that 'web + phone has got to be better' (ibid.)
24. According to the International Data Corporation, in 2010 a total of 302.6 million smartphones were sold, up 74.4% from 2009 (IDC 2011); the proportion of people with a smartphone at the end of 2010 was 27% in the U.S. and 31% in Europe (comScore 2011). A 2010 Forrester survey showed that 47% of mobile telephone users in Japan accessed the mobile Internet at least once a month, with China coming in at 43%, the U.S. at 22% and Europe at 12% (Reitsma 2011).
25. There are several location techniques available within the cellular network, such as triangulation and multilateration, that can help increase the accuracy of locating

mobile telephones, but there would still be an uncertainty in location of 100 metres or more.

26. In 2010, GPS-enabled devices amounted to 295 million units, with forecasts seeing that number rise to 940 million units by 2015 (Berg Insight 2011).
27. For a comprehensive taxonomy of location-based services, see Ratti et al. (2006: 729-731).
28. In the tenth century Blåtand brought rivalling tribes in Denmark and Norway together under one rule (Gow & Smith 2006: 125).
29. See for instance IBM's Smarter Planet website, where we learn that through instrumenting the world's systems and interconnecting them, all we have to do next is make them intelligent and the planet will be 'smart' and even know 'world peace' (IBM 2010). Otlet and Licklider would have been proud.
30. Sterling, speaking at the LIFT 2006 conference, noted as well that an Internet of Things will logically 'grow out of what we have right now', as 'a natural evolution from a world of digital devices that people are already carrying, laptops, media players, camera phones, WANs, and the wireless and wired land broadband networks that are serving them in locations as they move about, and the global Internet, and its growing store of socially generated knowledge, and Web-based on demand social applications' (Sterling 2006b).

References

Aakhus, Mark. 2003. Understanding information and communication technology and infrastructure in everyday life: Struggling with communication-at-a-distance. In *Machines that become us: The social context of personal communication technology*, edited by James E. Katz, 27-42. New Brunswick: Transaction Publishers.

Abanes, Richard. 1998. *End-time visions: The road to Armageddon?* New York: Four Walls Eight Windows.

Abramson, Albert. 1987. *The history of television, 1880 to 1941*. Jefferson, NC: McFarland.

Agar, Jon. 2003. *Constant touch: A brief history of the mobile phone*. Cambridge: Icon.

Ahonen, Tomi. 2006. They ARE the Borg: Youth, mobile and SMS text messaging. *Communities Dominate Brands Blog* http://communitiesdominate.blogs.com/brands/2006/03/they_are_the_bo.html.

Aitken, Hugh G.J. 1976. *Syntony and spark: The origins of radio*. New York: Wiley.

Aitken, William. 1939. *Who invented the telephone?* London/Glasgow: Blackie & Son.

Allen, Woody. 1972. *Play it again, Sam*. USA: Paramount Pictures.

Allison, Rebecca. 2003. Climbers on Alpine ridge rescued by text message. *The Guardian* http://www.guardian.co.uk/uk/2003/oct/07/travelnews.travel.

Allon, Fiona. 2004. An ontology of everyday control: Space, media flows and 'smart' living in the absolute present. In *Mediaspace: Place, scale, and culture in a media age*, edited by Nick Couldry and Anna McCarthy, 253-274. London/New York: Routledge.

Arnheim, Rudolf. 1957 [1935]. A forecast of television. In *Film as art*, edited by Rudolf Arnheim, 188-198. Berkeley: University of California Press.

Arnold, Michael. 2003. On the phenomenology of technology: The 'Janus-faces' of mobile phones. *Information and Organization* 13 (4): 231-256.

Aronson, Sidney H. 1977. Bell's electrical toy: What's the use? The sociology of early telephone usage. In *The social impact of the telephone*, edited by Ithiel de Sola Pool, 15–39. Cambridge, MA: MIT Press.

Atran, Scott. 2004. *In gods we trust: The evolutionary landscape of religion*. Oxford/New York: Oxford University Press.

Baczko, Bronislaw. 1978. *Lumières de l'utopie*. Paris: Payot.

Bailey, Lee Worth. 2005. *The enchantments of technology*. Urbana: University of Illinois Press.

Baldry, H. C. 1952. Who invented the Golden Age? *The Classical Quarterly* 2 (1/2): 83-92.

——. 1956. Hesiod's five ages. *Journal of the History of Ideas* 17 (4): 553-554.

Barlow, John Perry. 1996. A declaration of the independence of cyberspace. *Barlow Home (stead)Page* https://projects.eff.org/~barlow/Declaration-Final.html.

Baron, Naomi S. 2011. Concerns about mobile phones: A cross-national study. *First Monday* 16 (8) http://www.uic.edu/htbin/cgiwrap/bin/ojs/index.php/fm/article/view/3335/3032.

Barthes, Roland. 1972 [1957]. *Mythologies*. Translated by Annette Lavers. New York: Hill and Wang.

Basalla, George. 1988. *The evolution of technology*. Cambridge/New York: Cambridge University Press.

Baudrillard, Jean. 1993. Hyperreal America. *Economy and Society* 22 (2): 243-252.

BBC News. 2004. European press ask 'Why?' *BBC News International* http://news.bbc.co.uk/2/hi/europe/3504032.stm.

Berg Insight. 2011. Shipments of GPS-enabled GSM/WCDMA handsets grew 97 percent in 2010. *Berg Insight News* http://www.berginsight.com/News.aspx?m_m=6&s_m=1.

Bergson, Henri. 1910 [1889]. *Time and free will: An essay on the immediate data of consciousness*. Translated by Frank Lubecki Pogson. London/New York: Macmillan.

Bernstein, Richard J. 1971. *Praxis and action: Contemporary philosophies of human activity*. Philadelphia: University of Pennsylvania Press.

Bertho-Lavenir, Catherine. 1988. The telephone in France 1879 to 1979: National characteristics and international influcences. In *The development of large technical systems*, edited by Renate Mayntz and Thomas Parke Hughes, 69-104. Boulder: Westview Press.

Bijker, Wiebe E., and John Law, eds. 1992. *Shaping technology/building society: Studies in sociotechnical change*. Cambridge, MA: MIT Press.

Blaine, Robert Gordon. 1905. *Ætheric or wireless telegraphy*. London: Biggs & Sons.

Bloch, Ernst. 1959. *Das prinzip hoffnung*. Frankfurt am Main: Suhrkamp.

Bloom, Howard. 2000. *The global brain: The evolution of mass mind from the Big Bang to the 21st century*. New York: Wiley.

——. 2001. The omnologist manifesto. *Howard Bloom's Big Bang Tango Media Lab* http://www.bigbangtango.org/website/OmnologistManifesto.htm.

Blumler, Jay G., and Elihu Katz. 1974. *The uses of mass communications: Current perspectives on gratifications research*. Beverly Hills: Sage Publications.

Boddy, William. 1990. *Fifties television: The industry and its critics*. Urbana: University of Illinois Press.

——. 2004. *New media and popular imagination: Launching radio, television, and digital media in the United States*. Oxford/New York: Oxford University Press.

Boehm, Christopher. 1999. *Hierarchy in the forest: The evolution of egalitarian behavior*. Cambridge, MA: Harvard University Press.

Boettinger, Henry M. 1977. Our sixth-and-a-half sense. In *The social impact of the telephone*, edited by Ithiel de Sola Pool, 200-207. Cambridge, MA: MIT Press.

Bolter, Jay David, and Richard Grusin. 1999. *Remediation: Understanding new media*. Cambridge, MA: MIT Press.

Boomen, Marianne van den. 2007. Van gemeenschap via webnetwerk naar datawolk. In *Jaarboek ICT en samenleving 2007: Gewoon digitaal*, edited by Jan Steyaert and Jos de Haan, 129-148. Amsterdam: Boom.

——. 2012. *Transcoding the Internet: How metaphors matter in digital praxis*. Utrecht: University of Utrecht.

Borges, Jorge Luis. 1998. The library of Babel. In *Collected fictions*, edited by Jorge Luis Borges, 112-118. Translated by Andrew Hurley. New York: Viking.

Borgmann, Albert. 1999. *Holding on to reality: The nature of information at the turn of the millennium*. Chicago: University of Chicago Press.

Brecht, Bertolt. 1964 [1932]. The radio as an apparatus of communication. In *Brecht on theatre: The development of an aesthetic*, edited by Bertolt Brecht, 51-53. London: Methuen.

Briggs, Asa. 1977. The pleasure telephone: A chapter in the prehistory of media. In *The social impact of the telephone*, edited by Ithiel de Sola Pool, 40-65. Cambridge, MA: MIT Press.

Buckland, Michael. 1992. Emanuel Goldberg, electronic document retrieval, and Vannevar Bush's memex. *Journal of the American Society for Information Science* 43 (4): 284-294.

Bury, J. B. 1920. *The idea of progress: An inquiry into its origin and growth*. London/New York: Macmillan.

Bush, Vannevar. 1945. As we may think. *The Atlantic Monthly* 176 (1): 101-108.

Buskes, Chris. 2006. *Evolutionair denken: De invloed van Darwin op ons wereldbeeld*. Amsterdam: Nieuwezijds.

Cairncross, Frances. 1997. *The death of distance: How the communications revolution will change our lives*. Boston: Harvard Business School Press.

Campbell, Joseph. 1949. *The hero with a thousand faces*. New York: Pantheon Books.

——. 1959-1968. *The masks of God*. New York: Viking.

Campbell, Scott and Tracy Russo. 2003. The social construction of mobile telephony: An application of the social influence model to perceptions and uses of mobile phones within personal communication networks. *Communication Monographs* 70 (4): 317-334.

Campbell, Scott. 2007. A cross-cultural comparison of perceptions and uses of mobile telephony. *New Media & Society* 9 (2): 343-363.

Carey, James W. 1989. *Communication as culture: Essays on media and society*. Boston: Unwin Hyman.

Casson, Herbert Newton. 1910. *The history of the telephone*. Chicago: A.C. McClurg & Co.

Castells, Manuel. 1996. *The information age: Economy, society and culture, volume 1: The rise of the network society*. Oxford/Malden, MA: Blackwell.

Castells, Manuel, Mireia Fernandez-Ardevol, Jack Linchuan Qiu, and Araba Sey. 2007. *Mobile communication and society: A global perspective*. Cambridge, MA: MIT Press.

Catania, Basilio. 2003. Antonio Meucci, l'inventore del telefono. *Notiziario Tecnico Telecom Italia* 12 (1): 109-117.

Chandler, Daniel. 2002. Technological or media determinism: Introduction. *The Media and Communications Studies Site* http://www.aber.ac.uk/media/Documents/tecdet/tdet01.html.

Chaplin, Charles. 1936. *Modern times*. USA: United Artists.

Charny, Ben. 2003. Q&A: Mobile phone inventor says dream isn't real yet. *ZDNet.co.uk* http://www.zdnet.co.uk/news/desktop-hardware/2003/04/07/qanda-mobile-phone-inventor-says-dream-isnt-real-yet-2133081/.

Cherry, Colin. 1971. *World communication: Threat or promise? A socio-technical approach*. London/New York: Wiley-Interscience.

Chipchase, Jan, Per Persson, Petri Piipo, Mikko Aarras, and Tetsuya Yamamoto 2005. Mobile essentials: field study and concepting. *Proceedings of the 2005 conference on Designing for User eXperience*, paper 57. http://portal.acm.org/citation.cfm?id=1138301&CFID=38601465&CFTOKEN=29578949.

Churchill, Elizabeth F., and Nina Wakeford. 2002. Framing mobile collaborations and mobile technologies. In *Wireless world: Social and interactional aspects of the mobile age*, edited by Barry Brown, Nicola Green and Richard Harper, 154-179. London/New York: Springer.

CNN. 2008. Violence spreading in Kenya, Red Cross official says. *CNN International* http://edition.cnn.com/2008/WORLD/africa/01/28/kenya.violence/index.html.

Coe, Lewis. 1995. *The telephone and its several inventors: A history*. Jefferson, NC: McFarland.

Cohen, Stanley. 2002. *Folk devils and moral panics: The creation of the mods and rockers*. London/New York: Routledge.

Community Wireless. 2001. Your community: Online and wirefree. *Community Wireless* http://www.communitywireless.org/.

comScore. 2011. comScore releases inaugural report, 'The 2010 mobile year in review'. *comScore, Inc.* http://www.comscore.com/Press_Events/Press_Releases/2011/2/comScore_Releases_Inaugural_Report_The_2010_Mobile_Year_in_Review.

Crary, Jonathan. 1990. *Techniques of the observer: On vision and modernity in the nineteenth century*. Cambridge, MA: MIT Press.

Crombag, Hans, and Frank Van Dun. 1997. *De utopische verleiding*. Amsterdam: Contact.

Crookes, William. 1892. Some possibilities of electricity. *Fortnightly Review* 51: 173-181.

Cumiskey, Kathleen M. 2005. 'Can you hear me now?': Paradoxes of techno-intimacy resulting from the public use of mobile communication technology. In *A sense of place: The global and the local in mobile communication*, edited by János Kristóf Nyíri, 151-158. Vienna: Passagen.

Cusk, Rachel. 2001. Finding words. *The Guardian* http://www.guardian.co.uk/world/2001/sep/14/september11.politicsphilosophyandsociety1.

Czitrom, Daniel J. 1982. *Media and the American mind: From Morse to McLuhan*. Chapel Hill: University of North Carolina Press.

Davie, Ronald, Charlotte Panting, and Tony Charlton. 2004. Mobile phone ownership and usage among pre-adolescents. *Telematics and Informatics* 21 (4): 359-373.

Davis, Erik. 1998. *Techgnosis: Myth, magic, mysticism in the age of information*. New York: Harmony Books.

Davis, Robert E. 1976. *Response to innovation: A study of popular argument about new mass media*. New York: Arno Press.

Dawkins, Richard. 1976. *The selfish gene*. Oxford/New York: Oxford University Press.

——. 2006. *The God delusion*. Boston: Houghton Mifflin Company.

De Gournay, Chantal, and Zbigniew Smoreda. 2003. Communication technology and sociability: Between local ties and 'global ghetto'? In *Machines that become us: The social context of personal communication technology*, edited by James E. Katz, 57-70. New Brunswick: Transaction Publishers.

De Kerckhove, Derrick. 1997. *Connected intelligence: The arrival of the Web society*. Toronto: Somerville House.

Deleuze, Gilles, and Félix Guattari. 2004 [1988]. *A thousand plateaus: Capitalism and schizophrenia*. Translated by Brian Massumi. Minneapolis: University of Minnesota Press.

Dennett, Daniel C. 1995. *Darwin's dangerous idea: Evolution and the meanings of life*. New York: Simon & Schuster.

——. 2006. *Breaking the spell: Religion as a natural phenomenon*. New York: Viking.

Desaulniers, Jean-Pierre. 1986. Television and nationalism: From culture to communication. In *Television in transition: Papers from the first international television studies conference*, edited by Phillip Drummond and Richard Paterson, 112-122. London: BFI.

Desroche, Henri. 1979 [1973]. *The sociology of hope*. Translated by Carol Martin-Sperry. London/ Boston: Routledge & Kegan Paul.

DeVito, Joseph A. 1976. *The interpersonal communication book*. New York: Harper & Row.

Domhoff, G. William. 1996. *Finding meaning in dreams: A quantitative approach*. London/New York: Plenum.

——. 2005. The content of dreams: Methodologic and theoretical implications. In *Principles and practice of sleep medicine*, edited by Meir H. Kryger, Thomas Roth, and William C. Dement, 522-534. Philadelphia/Edinburgh: Elsevier Saunders.

Donner, Jonathan. 2003. What mobile phones mean to Rwandan entrepreneurs. In *Mobile democracy: Essays on society, self and politics*, edited by János Kristóf Nyíri, 393-410. Vienna: Passagen.

——. 2005. The mobile behaviours of Kigali's microentrepreneurs: Whom they call... And why. In *A sense of place: The global and the local in mobile communication*, edited by János Kristóf Nyíri, 293-304. Vienna: Passagen.

Douglas, Susan J. 1987. *Inventing American broadcasting, 1899-1922*. Baltimore: Johns Hopkins University Press.

Drummond, Kent, and Robert Hopper. 1991. Misunderstanding and its remedies: Telephone miscommunication. In *'Miscommunication' and problematic talk*, edited by Nikolas Coupland, Howard Giles, and John M. Wiemann, 301-314. Newbury Park: Sage Publications.

Du Boff, Richard B. 1984. The telegraph in nineteenth-century America: Technology and monopoly. *Comparative Studies in Society and History* 26 (4): 571-586.

Du Moncel, Theodore. 1879. *The telephone, the microphone, and the phonograph*. New York: Harper.

Dunlap, Orrin E. 1932. *The outlook for television*. New York: Harper & Brothers.

——. 1937. *Marconi: The man and his wireless*. London/New York: Macmillan.

Dussel, Enrique. 1993. Eurocentrism and modernity (introduction to the Frankfurt lectures). *boundary 2* 20 (3): 65-76.

Eco, Umberto. 1995. *The search for the perfect language*. Oxford: Blackwell.

Edwards, Paul N. 2003. Infrastructure and modernity: Force, time, and social organizations in the history of sociotechnical systems. In *Modernity and technology*, edited by Thomas J. Misa, Philip Brey and Andrew Feenberg, 185-225. Cambridge, MA: MIT Press.

Ellul, Jacques. 1964 [1954]. *The technological society*. Translated by John Wilkinson. New York: Knopf.

——. 1980 [1977]. *The technological system*. Translated by Joachim Neugroschel. New York: Continuum.

——. 1990 [1988]. *The technological bluff*. Translated by Geoffrey W. Bromiley. Grand Rapids: W.B. Eerdmans.

Elmore, Richard. 2005. Agency, reciprocity, and accountability in democratic education. In *The public schools*, edited by Susan Fuhrman and Marvin Lazerson, 277-301. Oxford/New York: Oxford University Press.

Elsner, Monika, Thomas Müller, and Peter M. Spangenberg. 1994. The early history of German television: The slow development of a fast medium. In *Materialities of communication*, edited by Hans Ulrich Gumbrecht and Karl Ludwig Pfeiffer, 107-143. Stanford: Stanford University Press.

Ericsson. 2006. Ericsson celebrates 50 years of mobile telephony. *Ericsson* http://www.ericsson.com/thecompany/press/releases/2006/10/1081237.

——. 2007. Chronological overview 1981-1990. *Ericsson* http://www.ericsson.com/ericsson/corpinfo/compfacts/history/years_leadership/1981-1990.shtml.

Eurostat. 2011. Mobile phone subscriptions (per 100 inhabitants). *Eurostat* http://epp.eurostat.ec.europa.eu/tgm/table.do?tab=table&init=1&language=en&pcode=tin00060&plugin=0.

Evenson, A. Edward. 2000. *The telephone patent conspiracy of 1876: The Elisha Gray-Alexander Bell controversy and its many players*. Jefferson, NC: McFarland.

Farley, Tom. 2006a. Mobile telephone history – page 4. *Privateline.com* http://www.privateline.com/PCS/history4.htm.

——. 2006b. Mobile telephone history – page 6. *Privateline.com* http://www.privateline.com/PCS/history6.htm.

——. 2006c. Mobile telephone history – page 7. *Privateline.com* http://www.privateline.com/PCS/history7.htm.

Fiedler, Leslie. 1996. *Tyranny of the normal: Essays on bioethics, theology & myth*. Boston: D.R. Godine.

Fischer, Claude S. 1992. *America calling: A social history of the telephone to 1940*. Berkeley: University of California Press.

Fleming, John Ambrose. 1902. *Waves and ripples in water, air, and æther*. London/New York: E. & J.B. Young & Co.

Flichy, Patrice. 1995 [1991]. *Dynamics of modern communication: The shaping and impact of new communication technologies*. Translated by Liz Libbrecht. London/Thousand Oaks: Sage Publications.

——. 2007 [2001]. *The Internet imaginaire*. Translated by Liz Carey-Libbrecht. Cambridge, MA: MIT Press.

Fortunati, Leopoldina. 2003. The mobile phone and democracy: An ambivalent relationship. In *Mobile democracy: Essays on society, self and politics*, edited by János Kristóf Nyíri, 239-258. Vienna: Passagen.

Foucault, Michel. 1972 [1969]. *The archaeology of knowledge*. Translated by Alan Sheridan-Smith. London: Tavistock Publications.

Fox, Barry. 1990. Confused over personal communications. *New Scientist* (1739): 44-48.

Fox, Kate. 2001. Evolution, alienation and gossip: The role of mobile telecommunications in the 21st century. *Social Issues Research Centre* http://www.sirc.org/publik/gossip.shtml.

Frankl, Viktor Emil. 1992 [1946]. *Man's search for meaning: An introduction to logotherapy*. Translated by Ilse Lasch. Boston: Beacon Press.

Friedman, Thomas L. 2005. *The world is flat: A brief history of the twenty-first century*. New York: Farrar, Straus and Giroux.

Frissen, Valerie. 1995. Gender is calling: Some reflections on past, present and future uses of the telephone. In *The gender-technology relation: Contemporary theory and research*, edited by Keith Grint and Rosalind Gill, 79-94. London: Taylor & Francis.

——. 2000. ICTs in the rush hour of life. *Information Society* 16 (1): 65-75.

Frost, Stanley. 1922. Radio dreams that can come true. *Collier's*, June 10, 18.

Geelen, Anne. 2006. *We all communicate. But do we understand each other?* Netherlands: VPRO.

Geoghegan, Vincent. 1996. *Ernst Bloch*. London/New York: Routledge.

——. 2007. Utopia and the memory of religion. In *Exploring the utopian impulse: Essays on utopian thought and practice*, edited by Michael Griffin and Tom Moylan, 101-116. Oxford: Peter Lang.

Gergen, Kenneth J. 2002. The challenge of absent presence. In *Perpetual contact: Mobile communication, private talk, public performance*, edited by James E. Katz and Mark Aakhus, 227-241. Cambridge/New York: Cambridge University Press.

——. 2003. Self and community in the new floating worlds. In *Mobile democracy: Essays on society, self and politics*, edited by János Kristóf Nyíri, 103-114. Vienna: Passagen.

Gershenfeld, Neil. 1999. *When things start to think*. New York: Henry Holt.

Geser, Hans. 2004. Towards a sociological theory of the mobile phone. *Sociology in Switzerland: Sociology of the Mobile Phone* http://socio.ch/mobile/t_geser1.pdf.

Gibson, William. 1984. *Neuromancer*. New York: Ace Books.

Gilbert, Alorie. 2005. Vegas casino bets on RFID. *CNET News.com* http://news.cnet.com/Vegas-casino-bets-on-RFID/2100-7355_3-5568288.html.

Ginsberg, Morris. 1953. *The idea of progress: A revaluation*. London: Methuen.

Gitelman, Lisa, and Geoffrey B. Pingree, eds. 2003. *New media, 1740-1915*. Cambridge, MA: MIT Press.

Goggin, Gerard. 2006. *Cell phone culture: Mobile technology in everyday life*. London/New York: Routledge.

Gombrich, Ernst. 1974. The Renaissance: Period or movement? In *Background to the English Renaissance: Introductory lectures*, edited by Arthur Geoffrey Dickens, 9-30. London: Gray-Mills Publishing.

Gordon, Eric, and Adriana de Souza e Silva. 2011. *Net locality: Why location matters in a networked world*. Boston: Blackwell-Wiley.

Gould, Stephen Jay. 1997. Evolution: The pleasures of pluralism debate. *The New York Review of Books* 44 (11): 47-52.

——. 2002. *The structure of evolutionary theory*. Cambridge, MA: Harvard University Press.

Gow, Gordon, and Richard Smith. 2006. *Mobile and wireless communications: An introduction*. Maidenhead: Open University Press.

Grameen Foundation. 2011. Empowering the poor. *Grameen Foundation* http://www.grameenfoundation.org/what-we-do/empowering-poor.

Green, Nicola. 2002. Who's watching whom? Monitoring and accountability in mobile relations. In *Wireless world: Social and interactional aspects of the mobile age*, edited by Barry Brown, Nicola Green and Richard Harper, 32-45. London/New York: Springer.

——. 2003. Outwardly mobile: Young people and mobile technologies. In *Machines that become us: The social context of personal communication technology*, edited by James E. Katz, 201-217. New Brunswick: Transaction Publishers.

Grint, Keith, and Steve Woolgar. 1997. *The machine at work: Technology, work, and organization*. Cambridge: Polity Press.

Groen, Gertjan. 2006. Geschiedenis van de mobiele telefonie. *Telecomwereld* http://www.telecomwereld.nl/mobgesch.htm.

Groys, Boris. 2002. On the new. *#ArtNodes* http://www.uoc.edu/artnodes/espai/eng/art/groys1002/groys1002.html.

GSM Association. 2006. GSM mobile phone technology adds another billion connections in just 30 months. *GSM World* http://www.gsmworld.com/newsroom/press-releases/2047.htm.

——. 2010. GSMA announces that global mobile connections surpass 5 billion. *GSM World* http://www.gsmworld.com/newsroom/press-releases/2010/5265.htm.

Gunkel, David. 2000. We are Borg: Cyborgs and the subject of communication. *Communication Theory* 10 (3): 332-357.

Gunning, Tom. 1991. Heard over the phone: The Lonely Villa and the de Lorde tradition of the terrors of technology. *Screen* 32 (2): 184-196.

Hall, Stuart. 1980. Encoding/decoding. In *Culture, media, language: Working papers in cultural studies*, 1972-79, edited by Stuart Hall, Dorothy Hobson, Andrew Lowe and Paul Willis, 128-138. London: Hutchinson.

Hall, Stuart, Dorothy Hobson, Andrew Lowe, and Paul Willis, eds. 1980. *Culture, media, language: Working papers in cultural studies*, 1972-79. London: Hutchinson.

Harman, Chris. 2004. The rise of capitalism. *International Socialism*, Issue 102 http://www.isj.org.uk/?id=21.

Hayles, N. Katherine. 2005. *My mother was a computer: Digital subjects and literary texts*. Chicago: University of Chicago Press.

Headrick, Daniel R. 2000. *When information came of age: Technologies of knowledge in the age of reason and revolution*, 1700-1850. Oxford/New York: Oxford University Press.

Hegel, Georg Wilhelm Friedrich. 1967 [1807]. *The phenomenology of mind*. Translated by James Black Baillie. New York: Harper & Row.

Heidegger, Martin. 1927. *Sein und Zeit*. Halle a.d. Saale: Niemeyer.

Heilbroner, Robert L. 1994. Do machines make history? In *Does technology drive history? The dilemma of technological determinism*, edited by Merritt Roe Smith and Leo Marx, 54-65. Cambridge, MA: MIT Press. Originally published in *Technology and Culture* 8 (July 1967): 335-345.

Highton, Edward. 1852. *The electric telegraph: Its history and progress*. London: J. Weale.

Hirsch, Eric. 1998. New technologies and domestic consumption. In *The television studies book*, edited by Christine Geraghty and David Lusted, 158-174. London/New York: Hodder Arnold.

Hofstadter, Douglas R. 1979. *Gödel, Escher, Bach: An eternal golden braid*. New York: Basic Books.

Holloway, Mark. 1966. *Heavens on earth: Utopian communities in America*, 1680-1880. New York: Dover Publications.

Holzmann, Gerard, and Björn Pehrson. 1994. *The early history of data networks*. Los Alamitos: IEEE Computer Society Press.

Honda. 2006. Museum/Asimo. TV advertisement directed by Peter Thwaites. UK: Wieden + Kennedy. http://www.honda.co.uk/asimo.

Horkheimer, Max, and Theodor W. Adorno. 1947. *Dialektik der Aufklärung: Philosophische Fragmente*. Amsterdam: Querido.

Horst, Heather A., and Daniel Miller. 2006. *The cell phone: An anthropology of communication*. New York: Berg.

Hughes, Thomas Parke. 1983. *Networks of power: Electrification in western society*, 1880-1930. Baltimore: Johns Hopkins University Press.

——. 2004. *Human-built world: How to think about technology and culture*. Chicago: University of Chicago Press.

Huhtamo, Erkki. 1994. From kaleidoscomaniac to cybernerd: Towards an archeology of the media. *De Balie Dossiers – Media Archaeology* http://www.debalie.nl/artikel.jsp?articleid=10104.

Huhtamo, Erkki, and Jussi Parikka. 2011. *Media archaeology: Approaches, applications, and implications*. Berkeley: University of California Press.

Hunt, Bruce. 1997. Doing science in a global empire: Cable telegraphy and electrical. Physics in Victorian Britain. In *Victorian science in context*, edited by Bernard V. Lightman, 312-333. Chicago: University of Chicago Press.

Hutchby, Ian. 2001. *Conversation and technology: From the telephone to the Internet*. Cambridge/ Malden, MA: Polity Press.

IBM. 2010. IBM – e-Government – Ideas – United States. *IBM Smarter Planet* http://www.ibm.com/smarterplanet/us/en/government/ideas.

IDC. 2011. Android rises, Symbian^3 and Windows Phone 7 launch as worldwide smartphone shipments increase 87.2% year over year, according to IDC. *International Data Corporation* http://www.idc.com/about/viewpressrelease.jsp?containerId=prUS22689111&sectionId=null&elementId=null&pageType=SYNOPSIS.

IEEE. 2008. IEEE milestones: One-way police radio communication, 1928. *IEEE Global History Network* http://www.ieeeghn.org/wiki/index.php/Milestones:One-Way_Police_Radio_Communication,_1928.

Inayatullah, Sohail. 2005. Framing the shapes and times of the future; towards a post-development vision of futures. *Metafutures.org* http://www.metafuture.org/sarkar/framing_the_future.htm.

Innis, Harold Adams. 1950. *Empire and communications*. Oxford: Clarendon Press.

ISC. 2011. Internet host count history. *Internet Systems Consortium, Inc.* http://www.isc.org/solutions/survey/history.

ITM, 2011. Global SMS traffic to reach 8.7 trillion in 2015. *Informa Telecoms & Media* http://www.informatm.com/itmgcontent/icoms/whats-new/20017843617.html.

Ito, Mizuko, Daisuke Okabe, and Misa Matsuda, eds. 2005. *Personal, portable, pedestrian: Mobile phones in Japanese life*. Cambridge, MA: MIT Press.

ITU. 2005a. ITU Internet reports 2005: The Internet of Things. *International Telecommunication Union* http://www.itu.int/osg/spu/publications/internetofthings.

——. 2005b. ITU Internet reports 2005: The Internet of Things – executive summary. *International Telecommunication Union* http://www.itu.int/osg/spu/publications/internetofthings/InternetofThings_summary.pdf.

——. 2010a. ITU estimates two billion people online by end 2010. *International Telecommunication Union* http://www.itu.int/net/pressoffice/press_releases/2010/39.aspx.

——. 2010b. Key Global Telecom Indicators for the World Telecommunication Service Sector. *International Telecommunication Union* http://www.itu.int/ITU-D/ict/statistics/at_glance/KeyTelecom.html.

Jamalipour, Abbas. 2003. *The wireless mobile Internet: Architectures, protocols, and services*. West Sussex/Hoboken: J. Wiley.

Jameson, Fredric. 2005. *Archaeologies of the future: The desire called Utopia and other science fictions*. New York: Verso.

Jaspers, Karl. 1955 [1935]. *Reason and existenz: Five lectures*. Translated by William Earle. New York: Noonday Press.

Jenkins, Henry. 1992. *Textual poachers: Television fans & participatory culture*. London/New York: Routledge.

——. 2006. *Fans, bloggers, and gamers: Exploring participatory culture*. New York: New York University Press.

Jenson, Scott. 2005. Default thinking: Why consumer products fail. In *The inside text: Social, cultural and design perspectives on sms*, edited by Richard Harper, Leysia Palen, and Alex Taylor, 305-326. Dordrecht: Springer.

Johnsen, Truls Erik. 2003. The social context of the mobile phone use of Norwegian teens. In *Machines that become us: The social context of personal communication technology*, edited by James E. Katz, 161-169. New Brunswick: Transaction Publishers.

Johnson, Steven. 2001. *Emergence: The connected lives of ants, brains, cities, and software*. New York: Scribner.

Jones, Julian. 2005. *How William Shatner changed the world*. USA: Allumination Filmworks.

Karabell, Zachary. 2001. *A visionary nation: Four centuries of the American dream and what lies ahead*. New York: HarperCollins.

Katz, James E. 2006. *Magic in the air: Mobile communication and the transformation of social life*. New Brunswick: Transaction Publishers.

Katz, James E., and Mark Aakhus, eds. 2002. *Perpetual contact: Mobile communication, private talk, public performance*. Cambridge/New York: Cambridge University Press.

Katz, James E., and Satomi Sugiyama. 2006. Mobile phones as fashion statements: Evidence from student surveys in the US and Japan. *New Media & Society* 8 (2): 321-337.

Kavanagh, Donncha, Kieran Keohane, and Carmen Kuhling. 2001. Reading Star Trek: Imagining, theorizing, and reflecting on organizational discourse and practice. In *Science fiction and organization*, edited by Warren Smith, Matthew Higgins, Martin Parker, and Geoff Lightfoot, 125-142. London/New York: Routledge.

Kellner, Douglas. 1997a. Ernst Bloch, utopia and ideology critique. In *Not yet: Reconsidering Ernst Bloch*, edited by Jamie Daniel and Tom Moylan, 80-95. London: Verso.

——. 1997b. From 1984 to one-dimensional man: Critical reflections on Orwell and Marcuse – section two. *Illuminations* http://www.uta.edu/huma/illuminations/kell13a.htm.

Kessler, Frank. 2002. *Het idee van vooruitgang in de mediageschiedschrijving*. Utrecht: University of Utrecht.

——. 2006. The cinema of attractions as *dispositif*. In *The cinema of attractions reloaded*, edited by Wanda Strauven, 57-69. Amsterdam: Amsterdam University Press.

Kierkegaard, Søren. 2006 [1843]. *Fear and trembling*. Translated by Sylvia Walsh. Cambridge/New York: Cambridge University Press.

King, John. 2007. Cell phones of dead still rang. *CNN.com – Anderson Cooper 360° blog* http://edition.cnn.com/CNN/Programs/anderson.cooper.360/blog/2007/04/volunteer-contributes-what-he-can-body.html.

Kingsbury, John E. 1915. *The telephone and telephone exchanges: Their invention and development*. London/New York: Longmans, Green & Co.

Kittler, Friedrich. 1999. *Gramophone, film, typewriter*. Stanford: Stanford University Press.

Kopomaa, Timo. 2000. *The city in your pocket: Birth of the mobile information society*. Helsinki: Gaudeamus.

Koszarski, Richard. 1998. Coming next week: Images of television in pre-war motion pictures. *Film History* 10 (2): 128-140.

Kürschner, Chris, Alexandra Brintrup, Paul Bowman, Erica Dubach, and Frédéric Thiesse 2010. Implementing RFID in Production Systems: A Case Study from a Confectionery Manufacturer. *Pacific Asia Journal of the Association for Information Systems* 2 (2): 19-42.

Kurzweil, Ray. 1999. *The age of spiritual machines: When computers exceed human intelligence*. New York: Viking.

Laland, Kevin N., and Gillian R. Brown. 2002. *Sense and nonsense: Evolutionary perspectives on human behaviour*. Oxford/New York: Oxford University Press.

Lang, Fritz. 1927. *Metropolis*. Germany: Paramount Pictures.

Lanier, Jaron. 2010. *You Are Not a Gadget: A Manifesto*. London/New York: Penguin Books.

Lardner, Dionysius, and Edward Brailsford Bright. 1867. *The electric telegraph*. London: J. Walton.

Larsen, Stephen, and Robin Larsen. 2002. *Joseph Campbell: A fire in the mind*. Rochester: Inner Traditions.

Lasswell, Harold D. 1938. *Propaganda technique in the world war*. New York: P. Smith.

——. 1948. The structure and function of communication in society. In *The communication of ideas: A series of addresses*, edited by Lyman Bryson, 37-51. New York: Institute for Religious and Social Studies.

Latour, Bruno. 1996. On actor-network theory. A few clarifications. *Soziale Welt* 47 (4): 369-381.

——. 2005. *Reassembling the social: An introduction to actor-network-theory*. Oxford/New York: Oxford University Press.

Leonardi, Paul, Marianne E. Leonardi, and Elizabeth Hudson. 2006. Culture, organization, and contradiction in the social construction of technology: Adoption and use of the cell phone across three cultures. In *The cell phone reader: Essays in social transformation*, edited by Anandam Kavoori and Noah Arceneaux, 205-225. New York: Peter Lang.

Levinson, Paul. 1997. *The soft edge: A natural history and future of the information revolution*. London/New York: Routledge.

——. 2004. *Cellphone: The story of the world's most mobile medium and how it transformed everything!* New York: Palgrave Macmillan.

Lévi-Strauss, Claude. 1963. *Structural anthropology*. Translated by Claire Jacobson and Brooke Grundfest Schoepf. New York: Basic Books.

——. 1987. *Anthropology and myth: Lectures, 1951-1982*. Translated by Roy Willis. Oxford/New York: Blackwell.

Lévy, Pierre. 1997 [1994]. *Collective intelligence: Mankind's emerging world in cyberspace*. Translated by Robert Bonomo. Cambridge, MA: Perseus Books.

——. 2005. Collective intelligence, a civilisation: Towards a method of positive interpretation. *International Journal of Politics, Culture, and Society* 18 (3-4): 189-198.

Licklider, Joseph Carl Robnett. 1960. Man-computer symbiosis. *IRE Transactions on Human Factors in Electronics* HFE-1: 4-11.

——. 1965. *Libraries of the future*. Cambridge, MA: MIT Press.

Licklider, Joseph Carl Robnett, and Robert W. Taylor. 1968. The computer as a communication device. *Science and Technology* (76): 21-31.

Licoppe, Christian. 2003. Two modes of maintaining interpersonal relations through telephone: From the domestic to the mobile phone. In *Machines that become us: The social context of personal communication technology*, edited by James E. Katz, 171-185. New Brunswick: Transaction Publishers.

Ling, Richard. 2004. *The mobile connection: The cell phone's impact on society*. San Francisco: Morgan Kaufmann.

Ling, Richard, and Per Helmersen. 2000. 'It must be necessary, it has to cover a need': The adoption of mobile telephony among pre-adolescents and adolescents. *Rich Ling home page* http://www.richardling.com/papers/2000_necesary_need.pdf.

Ling, Richard, and Birgitte Yttri. 2002. Hyper-coordination via mobile phones in Norway. In *Perpetual contact: Mobile communication, private talk, public performance*, edited by James E. Katz and Mark Aakhus, 139-169. Cambridge/New York: Cambridge University Press.

Lipartito, Kenneth. 2003. Picturephone and the information age: The social meaning of failure. *Technology and Culture* 44 (1): 50-81.

Lippmann, Walter. 1922. *Public opinion*. New York: Harcourt.

Luscombe, Stephen. 2007. Spanning the continents: Telegraphy. *British Empire: Science & Technology – Communications* http://www.britishempire.co.uk/science/communications/telegraph.htm.

Lynch, David. 1997. *Lost highway*. USA: October Films.

Lyon, David. 2003. *Surveillance as social sorting: Privacy, risk, and digital discrimination*. London/New York: Routledge.

MacKenzie, Donald. 1984. Marx and the machine. *Technology and Culture* 25 (3): 473-502.

Macy, Beverly, and Teri Thompson. 2011. *The Power of Real-Time Social Media Marketing: How to Attract and Retain Customers and Grow the Bottom Line in the Globally Connected World*. New York: McGraw-Hill Professional.

Mannheim, Karl. 1936 [1929]. *Ideology and utopia: An introduction to the sociology of knowledge*. Translated by Louis Wirth and Edward Shils. London/New York: Kegan Paul, Trench, Trubner & Co.

Manuel, Frank Edward, and Fritzie Prigohzy Manuel. 1979. *Utopian thought in the Western world*. Cambridge, MA: Belknap Press.

Marcuse, Herbert. 1964. *One-dimensional man: Studies in the ideology of advanced industrial society*. Boston: Beacon Press.

Marmasse, Natalia, and Chris Schmandt. 2003. Safe & sound: A wireless leash. Paper read at Conference on Human Factors in Computing Systems, in Fort Lauderdale, FL, USA.

Marvin, Carolyn. 1988. *When old technologies were new: Thinking about electric communication in the late nineteenth century*. Oxford/New York: Oxford University Press.

Marx, Karl, and Friedrich Engels. 1936 [1847]. *The poverty of philosophy*. Translated by Clemons Palme Dutt and V. Chattopadhyaya. London: Lawrence.

Marx, Leo. 1964. *The machine in the garden: Technology and the pastoral ideal in America*. Oxford/New York: Oxford University Press.

——. 1994. The idea of 'technology' and postmodern pessimism. In *Does technology drive history? The dilemma of technological determinism*, edited by Merritt Roe Smith and Leo Marx, 237-257. Cambridge, MA: MIT Press.

Mattelart, Armand. 2000 [1996]. *Networking the world, 1794-2000*. Translated by Liz Carey-Libbrecht and James A. Cohen. Minneapolis: University of Minnesota Press.

McGee, Michael C. 1980. The 'ideograph': A link between rhetoric and ideology. *The Quarterly Journal of Speech* 66 (1): 1-16.

McGuigan, Jim. 2005. Towards a sociology of the mobile phone. *Human Technology: An Interdisciplinary Journal on Humans in ICT Environments* 1 (1): 45-57.

Meulstee, Louis. 2007. Mobile radio in the Netherlands. *Wireless for the Warrior* http://wftw.nl/mobilophone/mobilophone.html.

Meyrowitz, Joshua. 1985. *No sense of place: The impact of electronic media on social behavior*. Oxford/New York: Oxford University Press.

——. 2003. Global nomads in the digital veldt. In *Mobile democracy: Essays on society, self and politics*, edited by János Kristóf Nyíri, 91-102. Vienna: Passagen.

Miller, Perry. 1961. The responsibility of mind in a civilization of machines. *The American Scholar* 31 (1): 51-69.

Misa, Thomas J. 1994. Retrieving sociotechnical change from technological determinism. In *Does technology drive history? The dilemma of technological determinism*, edited by Merritt Roe Smith and Leo Marx, 115-141. Cambridge, MA: MIT Press.

Mizrach, Steven. 2001. The symbolic invention of America-as-Utopia. *Utopian American Communities* http://www2.fiu.edu/~mizrachs/utopo-amer.html.

Moltmann, Jürgen. 1967. *Theology of hope: On the ground and the implications of a Christian eschatology*. New York: Harper & Row.

Mosco, Vincent. 2004. *The digital sublime: Myth, power, and cyberspace*. Cambridge, MA: MIT Press.

Motorola Inc. 2007. 1973-1983: Making history: Developing the portable cellular system. *Motorola Cell Phone Development* http://www.motorolasolutions.com/US-N/About/Company+Overview/History/Explore+Motorola+Heritage/Cell+Phone+Development.

Mul, Jos de. 2002. *Cyberspace odyssee*. Kampen: Klement.

Murray, Janet Horowitz. 1997. *Hamlet on the holodeck. The future of narrative in cyberspace*. New York: Free Press.

Museum voor Communicatie. 2006. Mobiele communicatie. *Museum voor Communicatie* http://www.muscom.nl/collecties/inhoud/artikel/083.htm.

Myerson, George. 2001. *Heidegger, Habermas and the mobile phone*. Duxford: Icon.

Nealon, Jeffrey T., and Caren Irr, eds. 2002. *Rethinking the Frankfurt School: Alternative legacies of cultural critique*. Albany: State University of New York Press.

Negley, Glenn Robert. 1977. *Utopian literature: A bibliography with a supplementary listing of works influential in utopian thought*. Lawrence: Regents Press of Kansas.

Negroponte, Nicholas. 1995. *Being digital*. New York: Knopf.

Newberg, Andrew B., Eugene G. D'Aquili, and Vince Rause. 2002. *Why God won't go away: Brain science and the biology of belief*. New York: Ballantine Books.

Newberg, Andrew B., and Mark Robert Waldman. 2006. *Why we believe what we believe: Uncovering our biological need for meaning, spirituality, and truth*. New York: Simon and Schuster.

Ng-Krülle, Grace, Paul Swatman, Felix Hampe, and Douglas Rebne. 2004. End-user perspectives on the adoption of wireless applications: Price of convenience and a model for contextual analysis. *ACIS 2004 Proceedings*, paper 46. http://aisel.aisnet.org/acis2004/46.

Nisbet, Robert A. 1980. *History of the idea of progress*. New York: Basic Books.

Noble, David F. 1997. *The religion of technology: The divinity of man and the spirit of invention*. New York: Knopf.

Nokia. 2006. It's what computers have become – the new Nokia N95. *Nokia Press Release* http://press.nokia.com/2006/09/26/its-what-computers-have-become-the-new-nokia-n95/.

NTT DoCoMo. 1999. Docomo vision 2010. *NTT DoCoMo press releases* http://www.nttdocomo.com/pr/1999/000868.html.

——. 2006. Toward 4G. *NTT DoCoMo future technologies* http://www.nttdocomo.com/technologies/future/toward/index.html.

——. 2010. Pursuing Smart Innovation: HEART. *NTT DoCoMo corporate vision* http://www.nttdocomo.com/about/core_foundation/corporate_vision/index.html.

Nye, David E. 1994. *American technological sublime*. Cambridge, MA: MIT Press.

NYPL. 2000. Bibliography of secondary sources. *New York Public Library* http://utopia.nypl.org/2ndsources.html.

Ogden, Charles Kay. 1935. *Basic English versus the artificial languages*. London: Kegan Paul, Trench, Trubner & Co.

Ogden, Charles Kay, and Ivor Armstrong Richards. 1956 [1923]. *The meaning of meaning: A study of the influence of language upon thought and of the science of symbolism*. London/Boston: Routledge & Kegan Paul.

O'Reilly, Tim. 2005. What is Web 2.0: Design patterns and business models for the next generation of software. *O'Reilly Media, Inc.* http://oreilly.com/web2/archive/what-is-web-20.html.

Orwell, George. 1951 [1949]. *Nineteen eighty-four: A novel*. London: Secker & Warburg.

Otlet, Paul. 1934. *Traité de documentation: Le livre sur le livre, théorie et pratique*. Brussels: Editiones Mundaneum.

Page, Benjamin. 1996. *Who deliberates? Mass media in modern democracy*. Chicago: University of Chicago Press.

Palen, Leysia, Marilyn Salzman, and Ed Youngs. 2000. Going wireless: Behavior & practice of new mobile phone users. Paper read at ACM Conference on Computer Supported Cooperative Work, in Philadelphia, PA, USA.

Paragas, Fernando. 2003. Dramatextism: Mobile telephony and people power in the Philippines. In *Mobile democracy: Essays on society, self and politics*, edited by János Kristóf Nyíri, 259-283. Vienna: Passagen.

Park, David W., Nicholas W. Jankowski, and Steve Jones. 2011. *The long history of new media: Technology, historiography, and contextualizing newness*. New York: Peter Lang.

Pearson, Carol, and Katherine Pope. 1981. *The female hero in American and British literature*. New York: R.R. Bowker.

Peters, John Durham. 1999. *Speaking into the air: A history of the idea of communication*. Chicago: University of Chicago Press.

Plant, Sadie. 2001. *On the mobile: The effects of mobile telephones on social and individual life*. Schaumburg: Motorola.

Plattel, Martin. 1970. *Utopie en kritisch denken*. Bilthoven: Ambo.

Pollard, Sidney. 1968. *The idea of progress: History and society*. London: Watts.

Pool, Ithiel de Sola, Craig Dekker, Stephen Dizard, Kay Israel, Pamela Rubin, and Barry Weinstein 1977. Foresight and hindsight: The case of the telephone. In *The social impact of the telephone*, edited by Ithiel de Sola Pool, 127-157. Cambridge, MA: MIT Press.

Pozner, Jennifer. 2000. In rape debate, controversy trumps credibility. *Fairness & Accuracy in Reporting* http://www.fair.org/index.php?page=1034.

Proust, Marcel. 1924 [1921]. *Remembrance of things past*. Translated by C. K. Scott Moncrieff. Vol. 2. New York: Random House.

Qiu, Jack Linchuan. 2007. The wireless leash: Mobile messaging service as a means of control. *International Journal of Communication* 1 (1): 74-91.

Rakow, Lana F., and Vija Navarro. 1993. Remote mothering and the parallel shift: Women meet the cellular phone. *Critical Studies in Mass Communication* 10: 144-157.

Ramati, Ayval. 1996. Harmony at a distance: Leibniz's scientific academies. *Isis* 87 (3): 430-452.

Rao, Bharat and Louis Minakakis. 2003. Evolution of mobile location-based services. *Communications of the ACM* 46 (12): 61-65.

Ratti, Carlo, Riccardo Maria Pulselli, Sarah Williams, and Dennis Frenchman 2006. Mobile landscapes: Using location data from cell phones for urban analysis. *Environment and Planning B: Planning and Design* 33 (5): 727-748.

Rayward, W. Boyd. 1975. *The universe of information: The work of Paul Otlet for documentation and international organisation*. Moscow: All-Union Institute for Scientific and Technical Information.

——. 1991. The case of Paul Otlet, pioneer of information science, internationalist, visionary: Reflections on biography. *Journal of Librarianship and Information Science* 23 (3): 135-145.

Reinhardt, Kurt Frank. 1952. *The existentialist revolt: The main themes and phases of existentialism*. Milwaukee: Bruce Publishing Company.

Reitsma, Reineke. 2011. The data digest: Urban China's mobile Internet use is double that of the US. *Forrester Blogs* http://blogs.forrester.com/reineke_reitsma/11-06-17-the_data_digest_urban_chinas_mobile_internet_use_is_double_that_of_the_us.

Relf, Jan. 1993. Utopia the good breast: Coming home to mother. In *Utopias and the Millennium*, edited by Krishan Kumar and Stephen Bann, 107-128. London: Reaktion Books.

Reuter, Michael. 1990. *Telekommunikation: Aus der Geschichte in die Zukunft*. Heidelberg: R. Decker.

Rheingold, Howard. 1993. *The virtual community: Homesteading on the electronic frontier*. Reading, MA: Addison-Wesley Publishing Company.

——. 2002. *Smart mobs: The next social revolution*. Cambridge, MA: Perseus Books.

Rice, Fredric L. 2001. A practical guide to the hero with a thousand faces by Joseph Campbell. *Scepticfiles.com* http://www.skepticfiles.org/atheist2/hero.htm.

Roberti, Mark. 2002. Trends: RFID. *CIO Insight* http://www.cioinsight.com/c/a/Trends/Trends-RFID.

Robida, Albert. 1883. *Le vingtième siècle*. Paris: G. Decaux.

Robison, David. 2003. Mobile privatization and the metaphors of mobile industries. In *Mobile democracy: Essays on society, self and politics*, edited by János Kristóf Nyíri, 165-176. Vienna: Passagen.

Rodowick, David. 1995. Audiovisual culture and interdisciplinary knowledge. *New Literary History* 26 (1): 111-121.

Roland, Alex. 1992. Theories and models of technological change: Semantics and substance. *Science, Technology & Human Values* 17 (1): 79-100.

Rowen, Robert. 2003. Gray and black radio propaganda against Nazi Germany. *New York Military Affairs Symposium* http://bobrowen.com/nymas/radioproppaper.htm.

Roy, Kaustuv. 2004. Overcoming nihilism: From communication to Deleuzian expression. *Educational Philosophy and Theory* 36 (3): 297-312.

Rule, James B. 2002. From mass society to perpetual contact: Models of communication technologies in social context. In *Perpetual contact: Mobile communication, private talk, public performance*, edited by James E. Katz and Mark Aakhus, 242-254. Cambridge/New York: Cambridge University Press.

Russell, Peter. 1995. *The global brain awakens: Our next evolutionary leap*. Palo Alto: Global Brain.

Sabine, Robert. 1867. *The electric telegraph*. London: Virtue Brothers & Co.

Sartre, Jean-Paul. 1956 [1943]. *Being and nothingness: An essay on phenomenological ontology*. Translated by Hazel Barnes. New York: Washington Square Press.

Schäfer, Mirko Tobias. 2008. Participation inside? User activities between design and appropriation. In *Digital material*, edited by Marianne van den Boomen et al., 147-158. Amsterdam: Amsterdam University Press.

——. 2011. *Bastard Culture! How user participation transforms cultural production*. Amsterdam: Amsterdam University Press.

Schehr, Robert C. 1997. *Dynamic utopia: Establishing intentional communities as a new social movement*. Oxford: Greenwood Publishing Group.

Schivelbusch, Wolfgang. 1986. *The railway journey: The industrialization of time and space in the 19th century*. Berkeley: University of California Press.

Schudson, Michael. 1978. The ideal of conversation in the study of mass media. *Communication Research* 5 (3): 320-329.

——. 1997. Why conversation is not the soul of democracy. *Critical Studies in Mass Communication* 14 (4): 297-309.

Sconce, Jeffrey. 2000. *Haunted media: Electronic presence from telegraphy to television*. Durham: Duke University Press.

Seadaniel. 2006. NTT Docomo vision 2010 part I-1 to 3. *Youtube* http://www.youtube.com/watch?v=eXuXBROyV-g, http://www.youtube.com/watch?v=zQZDhOoPfeI, http://www.youtube.com/watch?v=CqFkQswOoTE.

Shannon, Claude E. 1948. A mathematical theory of communication. *Bell System Technical Journal* 27: 379-423, 623-656.

——. 1949. *The mathematical theory of communication*. Urbana: University of Illinois Press.

Shaviro, Steven. 2003. *Connected: Or, what it means to live in the network society*. Minneapolis: University of Minnesota Press.

Shaw, William H. 1979. 'The handmill gives you the feudal lord': Marx's technological determinism. *History and Theory* 18 (2): 155-176.

Sherry, John, and Tony Salvador. 2002. Running and grimacing: The struggle for balance in mobile work. In *Wireless world: Social and interactional aspects of the mobile age*, edited by Barry Brown, Nicola Green and Richard Harper, 108-120. London/New York: Springer.
Shirky, Clay. 2009. *Here comes everybody: the power of organizing without organizations*. London/New York: Penguin Books.
——. 2010. *Cognitive surplus: Creativity and generosity in a connected age*. London: Penguin UK.
Sixma, Tjarda. 2007. Beeldbellen: De eeuwige belofte van 'nabijheid'. *Tijdschrift voor Mediageschiedenis* 10 (1): 88-10.
Skinner, B. F. 1948. *Walden Two*. London/New York: Macmillan.
——. 1981. Utopia as an experimental culture. In *America as Utopia*, edited by Kenneth M. Roemer, 28-42. New York: B. Franklin.
Skog, Berit. 2002. Mobiles and the Norwegian teen: Identity, gender and class. In *Perpetual contact: Mobile communication, private talk, public performance*, edited by James E. Katz and Mark Aakhus, 255-273. Cambridge/New York: Cambridge University Press.
Sloman, Edward. 1925. *Up the ladder*. USA: Universal Pictures.
Smith, Henry Nash. 1950. *Virgin land: The American West as symbol and myth*. Cambridge, MA: Harvard University Press.
Smith, Merritt Roe, and Leo Marx, eds. 1994. *Does technology drive history? The dilemma of technological determinism*. Cambridge, MA: MIT Press.
Smith, Michael L. 1994. Recourse of empire. In *Does technology drive history? The dilemma of technological determinism*, edited by Merritt Roe Smith and Leo Marx, 37-52. Cambridge, MA: MIT Press.
Snyder, C. Richard, ed. 2000. *Handbook of hope: Theory, measures and applications*. San Diego: Academic Press.
Sober, Elliott, and David Sloan Wilson. 1998. *Unto others: The evolution and psychology of unselfish behavior*. Cambridge, MA: Harvard University Press.
Sorensen, Roy. 2003. *A brief history of the paradox: Philosophy and the labyrinths of the mind*. Oxford/New York: Oxford University Press.
Spigel, Lynn. 1992. *Make room for tv: Television and the family ideal in postwar America*. Chicago: University of Chicago Press.
Standage, Tom. 1998. *The Victorian Internet: The remarkable story of the telegraph and the nineteenth century's on-line pioneers*. New York: Walker Publishing Company.
Stealth. 2008. Home. *Stealth Concealment Solutions, Inc.* http://www.stealthsite.com.
Steinbock, Dan. 2005. *The mobile revolution| The making of mobile services worldwide*. London/Sterling: Kogan Page.
Sterling, Bruce. 2005. *Shaping things*. Cambridge, MA: MIT Press.
——. 2006a. Media paleontology. In *Book of imaginary media: Excavating the dream of the ultimate communication medium*, edited by Eric Kluitenberg, 56-73. Rotterdam/London: NAi.
——. 2006b. Spimes and the future of artifacts. *LIFT 2006 Conference* http://liftconference.com/bruce-sterling-1.
Stewart, Jon. 2003. *Kierkegaard's relations to Hegel reconsidered*. Cambridge/New York: Cambridge University Press.
Stock, Brian. 2001. *After Augustine: The meditative reader and the text*. Philadelphia: University of Pennsylvania Press.

Symbian. 2007. Symbian fast facts q4 2007. *Symbian.com* http://www.symbian.com/about/fastfacts/fastfacts.html

T-Mobile. 2007. About T-Mobile hotspot. *T-Mobile* http://hotspot.t-mobile.com/services_about.htm.

——. 2010. Wi-fi hotspot for the way you live, work, and play. *T-Mobile* https://content.hotspot.t-mobile.com/AssetProcess.asp?asset=com.default.main.001.

Taylor, Alex, and Jane Vincent. 2005. An SMS history. In *Mobile world: Past, present, and future*, edited by Lynne Hamill and Amparo Lasen, 75-91. New York: Springer.

Taylor, Mark C. 2001. *The moment of complexity: Emerging network culture*. Chicago: University of Chicago Press.

Teilhard de Chardin, Pierre. 1959. *The phenomenon of man*. New York: Harper.

——. 1964. *The future of man*. New York: Harper & Row.

Telegram Service. 2008. Hoe werkt het □ telegram binnenland: Tarief. *Telegram Service BV* http://www.telegram.nl/ast.2.0/script/templet/tgs_1/template.asp?pageid=9#P3.

Tenner, Edward. 1996. *Why things bite back: Technology and the revenge of unintended consequences*. New York: Knopf.

The Cloud. 2007. Wireless solutions. *The Cloud* http://www.thecloud.net.

Theall, Donald. 2006. Understanding me: Lectures and interviews by Marshall McLuhan. *Canadian Journal of Communication* 30 (4): 695-698.

Thiagobassolli. 2007. NTT Docomo vision 2010. *Youtube* http://www.youtube.com/watch?v=ae-Ssclu5A4.

Thiesse, Frédéric, Christian Floerkemeier, Mark Harrison, Florian Michahelles, and Christof Roduner 2009. Technology, standards, and real-world deployments of the EPC network. *IEEE Internet Computing* 13 (2): 36-43.

Thomas, Frank. 1988. The politics of growth: The German telephone system. In *The development of large technical systems*, edited by Renate Mayntz and Thomas Parke Hughes, 179-214. Boulder: Westview Press.

Thompson, Henrietta. 2005. *Phone book: A handy guide to the world's favourite invention*. London: Thames & Hudson.

Thompson, Silvanus Phillips. 1883. *Philipp Reis: Inventor of the telephone*. London/New York: E. & F. N. Spon.

Thompson, William Irwin. 2004. *Self and society: Studies in the evolution of consciousness*. Charlottesville: Imprint Academic.

Thornhill, Randy, and Craig Palmer. 2000. *A natural history of rape: Biological bases of sexual coercion*. Cambridge, MA: MIT Press.

Townsend, Anthony M. 2002. Mobile communications in the twenty-first century city. In *Wireless world: Social and interactional aspects of the mobile age*, edited by Barry Brown, Nicola Green and Richard Harper, 62-77. London/New York: Springer.

Turrettini, Emily. 2004. Lost hikers use text messaging to gain help. *Textually.org* http://www.textually.org/textually/archives/2004/07/004683.htm.

——. 2007. Cell phones cause insomnia. *Textually.org* http://www.textually.org/textually/archives/2007/08/016895.htm.

Tuveson, Ernest Lee. 1949. *Milennium and utopia: A study in the background of the idea of progress*. Berkeley: University of California Press.

Uricchio, William. 2004. Storage, simultaneity, and the media technologies of modernity. In *Allegories of communication: Intermedial concerns from cinema to the digital*, edited by John Fullerton and Jan Olsson, 123-138. Rome: J. Libbey.

U.S. House of Representatives. 2002. H. Res. 269. *Library of Congress* http://frwebgate.access.gpo.gov/cgibin/getdoc.cgi?dbname=107_cong_bills&docid=f:hr269eh.txt.pdf.

Van Doren, Charles. 1967. *The idea of progress*. New York: F.A. Praeger.

Vattimo, Gianni. 1992 [1989]. *The transparent society*. Translated by David Webb. Baltimore: Johns Hopkins University Press.

Vincent, Jane. 2003. Emotion and mobile phones. In *Mobile democracy: Essays on society, self and politics*, edited by János Kristóf Nyíri, 215-224. Vienna: Passagen.

Virilio, Paul. 1997 [1995]. *Open sky*. Translated by Julie Rose. London/New York: Verso.

Vodafone UK. 2007. Raining time. TV advertisement directed by Simon Ratigan. UK: BBH. http://www.hla.net/GetMedia.php?id=138.

Vries, Imar de. 2006. The mobile phone: An ever-changing manifestation of the unchangeable. Paper read at Paris/Utrecht Mini Symposium, at Paris 8.

——. 2008. The vanishing points of mobile communication. In *Digital material*, edited by Marianne van den Boomen et al., 81-93. Amsterdam: Amsterdam University Press.

——. 2012. Through the looking cellphone screen: Dreams of omniscience in an age of mobile augmented reality. In *The mobile media reader*, edited by Noah Arceneaux & Anandam Kavoori, forthcoming. New York: Peter Lang.

Watson, Thomas Augustus. 1926. *Exploring life*. London/New York: D. Appleton & Co.

Wegner, Phillip E. 2002. *Imaginary communities: Utopia, the nation, and the spatial histories of modernity*. Berkeley: University of California Press.

Weinrich, Harald. 2004. *Lethe: The art and critique of forgetting*. Ithaca: Cornell University Press.

Wellman, Barry. 2002. Little boxes, glocalization, and networked individualism. In *Digital cities II: Computational and sociological approaches*, edited by Makoto Tanabe, Peter van den Besselaar, and Toru Ishida, 10-25. Berlin/New York: Springer.

Wells, H.G. 1938. *World brain*. Garden City: Doubleday, Doran & Co.

Wertheim, Margaret. 1999. *The pearly gates of cyberspace: A history of space from Dante to the Internet*. New York: W.W. Norton.

White, Thomas H. 2005. Military automobile for wireless telegraphy (1901). *United States Early Radio History* http://earlyradiohistory.us/1901auto.htm.

——. 2007. Personal communication by wireless (1879-1922). *United States Early Radio History* http://earlyradiohistory.us/sec004.htm.

Whyte, Jennifer. 2007. Evolutionary theories and design practices. *Design Issues* 23 (2): 46-54.

Wiener, Philip P. 1973. *Dictionary of the history of ideas: Studies of selected pivotal ideas*. New York: Scribner.

Wikipedia. 2011. Religions by country. *Wikipedia* http://en.wikipedia.org/wiki/Religions_by_country.

Wilde, Oscar. 1891. *The soul of man under socialism*. New York: Humboldt Publishing Company.

Wilde, Rein de. 2000. *De voorspellers: Een kritiek op de toekomstindustrie*. Amsterdam: De Balie.

Winston, Brian. 1998. *Media technology and society: A history from the telegraph to the Internet.* London/New York: Routledge.

Wit, Onno de. 1998. *Telefonie in Nederland 1877-1940: Opkomst en ontwikkeling van een grootschalig technisch systeem.* Amsterdam: O. Cramwinckel.

Wolfe, Tom. 2004. McLuhan's new world. *Wilson Quarterly* 28 (2): 18-25.

Wolpert, Lewis. 2006. *Six impossible things before breakfast: The evolutionary origins of belief.* London/New York: W.W. Norton & Co.

Wong, Paul. 2001. Logotherapy. *International Network on Personal Meaning* http://www.meaning.ca/archives/archive/art_logotherapy_P_Wong.htm.

Woolgar, Steve. 1991. Configuring the user: The case of usability trials. In *A sociology of monsters: Essays on power, technology and domination*, edited by John Law, 58-99. London/New York: Routledge.

Wright, Robert. 2000. *Nonzero: The logic of human destiny.* New York: Pantheon Books.

Wright, Ronald. 2004. *A short history of progress.* Toronto: House of Anansi Press.

Wu, Tim. 2010. *The master switch: the rise and fall of information empires.* New York: Alfred A. Knopf.

Young, Paul. 2003. Media on display: A telegraphic history of early American cinema. In *New media, 1740-1915*, edited by Lisa Gitelman and Geoffrey B. Pingree, 229-264. Cambridge, MA: MIT Press.

Zamyatin, Yevgeny Ivanovich. 1972 [1920]. *We.* Translated by Mirra Ginsburg. New York: EOS HarperCollins.

Zielinski, Siegfried. 1999 [1989]. *Audiovisions: Cinema and television as entr'actes in history.* Translated by Gloria Custance. Amsterdam: Amsterdam University Press.

——. 2006a [2002]. *Deep time of the media: Toward an archaeology of hearing and seeing by technical means.* Translated by Gloria Custance. Cambridge, MA: MIT Press.

——. 2006b. Modelling media for Ignatius Loyola. In *Book of imaginary media: Excavating the dream of the ultimate communication medium*, edited by Eric Kluitenberg, 28-55. Rotterdam/London: NAi.

Zimmer, Michael. 2008. The externalities of search 2.0: The emerging privacy threats when the drive for the perfect search engine meets Web 2.0. *First Monday* 13 (1) http://www.uic.edu/htbin/cgiwrap/bin/ojs/index.php/fm/article/view/2136/1944.

Zipes, Jack. 1988. Introduction: Toward a realization of anticipatory illumination. In *The utopian function of art and literature: Selected essays*, edited by Ernst Bloch, xi-xliii. Cambridge, MA: MIT Press.

Index

D

E

I

J

K

L

M

N

T

U

V

W

Y

Z

Other titles in the MediaMatters series

Bay-Cheng, S., C. Kattenbelt, A. Lavender, R. Nelson (eds.), *Mapping Intermediality in Performance*, 2010 (ISBN 978 90 8964 255 4)

Bleeker, M. (ed.), *Anatomy Live. Performance and the Operating Theatre*, 2008 (ISBN 978 90 5356 516 2)

Boomen, M. van den, S. Lammes, A. Lehmann, J. Raessens, M.T. Schäfer (eds.) *Digital Material. Tracing New Media in Everyday Life and Technology*, 2009 (ISBN 978 90 8964 068 0)

Lauwaert, M., *The Place of Play. Toys and Digital Cultures*, 2009 (ISBN 978 90 8964 080 2)

Schäfer, M.T., *Bastard Culture! How User Participation Transforms Cultural Production*, 2011 (ISBN 978 90 8964 256 1)

Sihvonen, T., *Players Unleashed! Modding The Sims and the Culture of Gaming*, 2011 (ISBN 978 90 8964 201 1)